SYSTEMS SCIENCES AND MODELLING

Trends in Scientific Research 1

SYSTEMS SCIENCES AND MODELLING

Edited by

A. RUBERTI
University of Rome, Italy

D. REIDEL PUBLISHING COMPANY
A MEMBER OF THE KLUWER ACADEMIC PUBLISHERS GROUP
DORDRECHT / BOSTON / LANCASTER

UNESCO, PARIS

Library of Congress Cataloging in Publication Data
Main entry under title:

CIP

Systems sciences and modelling.

Includes index.
1. System theory–Addresses, essays, lectures. 2. Mathematical models–Addresses, essays, lectures. I. Ruberti, Antonio.
Q295.S965 1984 003 84-8319
Reidel ISBN 90-277-1661-7 (hardbound)
Unesco ISBN 92-3-102138-9 (limpbound)

Published by The United Nations Educational,
Scientific and Cultural Organization,
7 Place de Fontenoy, 75700 Paris, France
and D. Reidel Publishing Company,
P.O. Box 17, 3300 AA Dordrecht, Holland

Hardbound edition sold and distributed in the U.S.A. and Canada
by Kluwer Academic Publishers, 190 Old Derby
Street, Hingham, MA 02043, U.S.A.

Hardbound edition in all other countries, sold and distributed
by Kluwer Academic Publishers Group,
P.O. Box 322, 3300 AH Dordrecht, Holland

D. Reidel Publishing Company is a member of the Kluwer Group.

Limpbound edition sold throughout the world by
Unesco, 7 Place de Fontenoy, 75700 Paris, France

Printed in The Netherlands

Contents

Preface

Unesco, in cooperation with the International Council of Scientific Unions (ICSU), and a number of international and national institutions, such as the International Federation of Institutes for Advanced Study, is publishing a series of monographs 'Trends in Scientific Research'. Each monograph will be complete in itself, but will be linked to the others by a central theme – that of developments at the forefront of science of special interest to humankind.

The idea is far from new. Professor P. Auger published a major work in 1957, 'Trends in Scientific Research', which played a seminal role in stimulating the development of science in many countries and served for many years as a handbook for science policy makers. Since this book was published, there has been a considerable extension of our scientific knowledge and it is no longer possible to fit all of science into one book if the various disciplines are to be treated in some depth. It was therefore decided to publish a number of separate monographs written by specialists and linked together to provide an extensive and extending image of present and future trends in science.

The monographs will be interdisciplinary in character, bringing together not only research in several disciplines but also the applied as well as the basic aspects of science, and in particular the uses of science in forecasting and in problem solving. Special attention will be paid to the use of science in the fulfilment of human needs.

The monographs are addressed to decision makers, to scientists interested in disciplines other than their own, the educated man in the street, and so on. It is intended that the series will provide interesting material that will influence young students to take up a career in the subjects treated.

Inasmuch as there will be a focus on applied aspects of the subjects treated, the monographs will also be of use in highlighting those types of scientific research most relevant to the requirements of developing countries and those fields of research and development with the greatest potential applications. Moreover, they will also try to highlight and foresee the most important aspects of the consequences on society of the introduction of new technologies.

The monographs will provide a series of interesting and stimulating publications which, it is hoped, will play an important role in shaping the development of scientific research and stimulating a greater interest in science in the world at large.

THE EDITORIAL BOARD

List of Contributors

Aoki, M., Professor, Institute of Social and Economic Research, Osaka University, Osaka, Japan.
Cohen, G., Director of the 'Centre d'Automatique et Informatique de l'Ecole Nationale Supérieure des Mines de Paris', Fontainebleau, France.
De Julio, S., Professor, Università della Calabria, Cosenza, Italy.
Gvishiani, J. M., Academician, Director of the Institute of Systems Studies, Moscow, U.S.S.R.
Guardabassi, G., Professor, Politecnico di Milano, Milan, Italy.
Rinaldi, S., Professor, Politecnico di Milano, Milan, Italy.
Ruberti, A., Professor, President of the Università di Roma, Italy.
Willems, J. C., Professor, University of Groningen, The Netherlands.

Introduction

Systems theory has grown to be a conceptually autonomous discipline over the last twenty-five years, out of the possibility of – and need for – a formalization and generalization of some advanced techniques in engineering design (particularly control engineering), which were made feasible by the introduction of computer technology.

As with most scientific disciplines, systems theory made use of several mathematical techniques, particularly in the field of differential equations and algebra, and provided a new impetus to new fields of mathematical research.

The validity of the use of the systems theoretic approach has, however, not remained confined to engineering alone, but has proved useful both in providing a refreshing unifying approach and in suggesting new methods in other disciplines such as economics, biology, the social sciences and physics.

The aim of this monograph is to outline the most promising trends of current research in the aspects of systems theory which deal with modelling and, partially, with controlling social, environmental and physical systems. In doing so, it may prove a useful tool for scientific policy makers in their decisions based on forecasts of research developments in the above fields, and for young graduates in the choice of their PhD thesis and the field of their future research. Moreover, it may be of interest to researchers in the social and physical sciences concerned with the systems theoretic point of view.

The above aim could be pursued only by using the expertise of an international team of systems theorists concerned with the application of systems theory to engineering, the social sciences, economics, environmental and physical sciences.

The level of the mathematical techniques which are used has been limited to a minimum in order to reach a wider audience, even if it has been unavoidable to use some mathematical language – mainly in the field of differential equations – in the articles on large-scale systems, dynamic economic phenomena and physics.

The first and preliminary article, written by S. De Julio and A. Ruberti, gives the basic conceptual foundations of systems theory, with particular reference to the process of building mathematical models from inputs and outputs, and of identifying a system. Moreover, it discusses critically the validity and reliability of the above quantitative techniques in dealing with large-scale social, economic and environmental systems, with a view to forecasting their behaviour and determining optimal or suboptimal control laws as inputs to their systems.

The following article by Academician J. M. Gvishiani, is of a qualitative-descriptive nature and mainly discusses the systems approach to the study of complex systems of various natures, and the solution which it may offer to multiple-factor and multiple-criteria problems. A distinction is made between formalizable and hardly-formalizable properties of systems, and the need to take into account 'human factors' in an organic unity with economic and technical factors is stressed. It also gives a rather detailed description of the "Modelling of global development" project which has been carried out since 1977 in the USSR.

A complete detailed overall mathematical model of some engineering systems, and most social, economic and environmental systems, cannot be profitably built using only a limited intellectual and/or computational power. These complex, real systems are called 'large-scale' systems. For these, either a coarser, more aggregated model is used, or the original system is decomposed into a set of interacting subsystems, of which the modelling and control is more readily feasible. Two articles are devoted to present research trends in the above areas, including their mathematical and computational aspects, as well as some engineering applications. More specifically, G. Guardabassi deals with the analysis of interconnected systems, the model of which consists of interacting submodels. The possibility of their computer simulation and their stability are discussed in a rather detailed way, taking into account the importance of these problems in the framework of Very-Large-Scale Integration (VLSI) technology for digital information processing. G. Cohen deals with designing optimal control of large-scale systems, via ad-hoc, or standard optimization techniques. Two methods of design are specifically described, the decomposition-coordination method and the decentralized control method. Reference is made to a particular type of large-scale system, namely power and water supply networks.

The article by S. Rinaldi deals with present trends of research in environmental sciences and the role played by systems sciences in simulating, planning, forecasting and managing water and land resources, air pollution, water pollution, ecological system and renewable resources.

Recent utilization of systems and control theoretic concepts and techniques in modelling economic problems in a dynamic framework are described by M. Aoki, who also gives an illustrative account of state space representations of dynamic problems drawn from time series models and policy models of international economics. He also shows how the state space method supplements other techniques common in economic theory, and enhances our ability to master complex dynamic phenomena in economic intertemporal optimization problems which would not otherwise render this feasible.

The final article, by J. C. Willems, discusses critically how a scientific discipline such as systems theory, which has crystallized out of the prescriptive aspects of certain advanced fields of engineering design, can be used in a typically descriptive process like modelling physical systems. In particular, it deals with the need for modelling physical systems with internal (state) and external variables, and with a system theoretic axiomatic basis for such models. Examples are drawn from classical thermodynamics and Hamilton mechanics.

A. RUBERTI

S. De Julio and A. Ruberti

General Methodologies in Systems Sciences and Mathematical Modelling

SCIENCE AND MODELS

Science, in all its stages, has always adopted some kind of model to interpret natural phenomena or artificial processes, so much so that the history of science could almost be said to coincide with that of models.

The main characteristics of the models built up until the present are their interpretative nature and quantitative formulation, and these are apparent in most processes of model building.

The *interpretative nature* comes from the need to know the phenomena under consideration better, and it is based on the adoption of physical laws or, more generally speaking, on theories by which these same phenomena can be explained. The so-called analogical models can also be related to this area of interpretation, having been created by discovering behavioural analogies among different phenomena, such that an already known phenomenon may be used as a model for another.

The *quantitative formulation* enables the mathematical construction of models which allow one to forecast the future evolution of a process or a phenomenon, once the initial conditions or causes, or the stimuli which are acting on it, are known. The characteristics that presently define the process of model construction may be recognized by considering recent trends in science evolution. One of the first trends was to extend quantitative analysis to any phenomenon, either physical or biological, economical or social. Another was to develop man's ability to predict and control phenomena, rather than just interpreting them. The increasing interest in the behaviour rather more than in the explanation of phenomena is partly a cause and partly a result of these two trends.[1]

These trends are naturally linked to the development of systems theory. In the context of systems theory a phenomenon (or process, or object) is considered in isolation from the 'universe' of which it is part and to which it is linked via different quantities that are classified in the following way:

– quantities which can be modified either directly or indirectly and that allow one to act on the phenomenon itself: these are defined as *input quantities* (*or inputs*) and can also be interpreted as control quantities;

– quantities which can be observed directly or indirectly: these are defined as *output quantities* (*or outputs*) and can be interpreted as controlled quantities;

– other quantities corresponding to interactions which cannot be ignored and somehow make up unchangeable inputs; they therefore usually take on the characteristics of *disturbances* because of the undesired effect which they can produce on the outputs.

The input and output quantities are associated with the corresponding time evolution. The set of all the possible input-output pairs is called an (*abstract*) *system* which is associated with the phenomenon (or process, or object) under consideration.

If we define it in this way, *the system appears to be usable as a general concept of a mathematical model* of a certain phenomenon and therefore can be potentially used with the widest range of goals and in the widest range of sectors. The generality of the definition indeed frees the system from the restrictions of particular disciplines and allows its greatest use. The interactions with the universe being based on the distinction between inputs and outputs, the definition also puts the prediction and control problems into immediate relief. As far as prediction is concerned, it is a question of finding the output corresponding to a given input. With the control, it is a question of determining the input corresponding to a desired output.

Finally, the complete *abstraction* with respect to the internal mechanisms that establish the dependencies between inputs and outputs allows the interpretative approach to be superseded and a behaviouristic approach used instead.

Apart from the width of scope of this approach, there has to be stressed the degree of arbitrariness linked to the isolation process of phenomena from the universe and to the evaluation of the interactions. In particular, the arbitrariness of the definition of inputs and outputs, i.e. of those features of the phenomenon which are held to be relevant and which are consistently linked to a specific point of view, should be noted. In any case, this has always taken place in the construction of models and the consequence is that *several models may correspond to a single phenomenon*. This is confirmation that the model can only represent partial aspects of a given phenomenon. A perfect and comprehensive model is not necessary because, even if one could be found, such a model would have the capacity of a direct understanding of the reality in its entirety [1].

Furthermore, it should be noted that the systems approach to the construction of models also brings in the directionality of links between inputs and outputs that some authors consider as an implicit use of the causality principle, on which, as is well known, positions can be quite different.

GENERAL METHODOLOGIES FOR BUILDING MATHEMATICAL MODELS

We have seen how the definition of the systems approach is based on the abstract system; it is, in fact, made up of all the possible input-output pairs that can be associated with a given phenomenon. At a high enough level of generalization, therefore, a system is a mathematical entity, since it is nothing more than a relationship between the set of all possible input functions and the set of all possible output functions. This entity can have different representations, from the explicit (a list of all the possible input-output pairs) to the implicit (via the mathematical conditions which the elements of each input-output pair must satisfy). Knowledge of the explicit representation of a system trivially resolves the prediction and control problems by picking out from the list all the pairs corresponding to a fixed input or output function respectively. It is clear that this path is hardly practicable, with the exception of very special cases. In order to resolve problems of prediction and control with implicit representation, we must use various mathematical theories and techniques, and (apart from elementary cases) make use of computers once adequate computing algorithms have been developed. It should be noted that the term mathematical model is often used for implicit representation rather than for the explicit one, even if both convey the same information, and this use of the word is therefore questionable. Nevertheless, given the fact that a list of input-output pairs is rarely available, it seems realistic to interpret the term mathematical model as *any implicit representation of an abstract system*. In this sense the connection between the use of models and the use of computers also becomes clear.

It is a fact that the use of models (as we interpret them here) requires calculations that are usually complex and/or heavy, and which therefore can never be carried out without the use of computers.[2] This explains why the development of model construction has grown in parallel with the evolution of the computer and computer science in general.

Having established these facts we can now move on to describe synthetically the construction methods of mathematical models.

Given a phenomenon, suppose that the process of isolation and selection has been performed; suppose, therefore, that the relative inputs and outputs have been defined. The selection of the abstract system which is to be associated with the phenomenon under examination takes the name *identification*. The phenomenon appears as a 'black box' and only a number of quantities, namely inputs and outputs, appear explicitly. The most direct identification process is obviously that of carrying out experiments designed to discover all the possible input-output pairs. The number of experiments which should be carried out is usually infinite[3]. This procedure is therefore impracticable. However, we cannot presume that a different approach to the identification problem will produce some simplification unless we also adopt extra hypotheses. In fact, the difficulties are intrinsic to the nature of the problem, which consists of picking out from the universe of systems the one corresponding to the case under consideration. We therefore have to limit the universe of the systems by splitting the problem of identification into two successive phases. In the first phase, we assume *a priori* that the system to be identified has particular properties, in such a way as to restrict the class of mathematical entities to which it belongs. In the second phase, we proceed to identification within this class.

As far as the first phase – which can be described as the identification of the mathematical models class – is concerned, two paths can be followed:

(a) *to assume a-priori hypotheses of a strictly mathematical kind*: for example, the hypothesis that the system could admit, as an implicit representation, a system of linear ordinary differential equations (linearity hypothesis) with constant coefficients (stationarity hypothesis). This assumption could be explained by the fact that the model of interest is valid locally (as far as excursions of variables are concerned) and by the fact that the time domain for which the model is to be used is such that the phenomenon characteristics (equations coefficients) can be considered invariant.

(b) *to give a rational interpretation of the behaviour mechanisms of the process*: thus the selection of an interpretative model comes out as an important stage in identification, even though it is used to restrict the class of models over which the identification procedure is applied. We speak of a process which tends to make the black box transparent.

The two paths are not mutually exclusive. On the contrary, they can be integrated to give a wide range of possibilities corresponding to the transfer from the black to the transparent box through a series of greys. Other terms used are identification from the outside (path a) and identification from the inside (path b). Others are reported as deductive and inductive processes: different degrees of induction depending on the different shades of grey which can be found between the limits of the black box and the transparent one [2].

THE INTERDISCIPLINARY APPROACH TO BUILDING MATHEMATICAL MODELS

Many people argue about the rising interest in model building and the increasing research activity in the area, involving both methodologies and applications. Many opponents of the systems approach base their arguments on the unsuccessful stages which, in fact, are typical of any new field of science during its adolescence. We will try to analyse, albeit briefly, possible misunderstandings that may have produced such difficulties.

Some Observations on the Systems Approach

If what we have been noting so far is true, it is clear that the systems approach is somehow the foundation of the present work on modelling. The fact that such an approach has been adopted in widely different areas clearly emphasises its unifying role, but at the same time it underlines the dangers brought in. Precisely because of the unifying possibilities of systems science with respect to problems, it is reasonably simple to reinterprete well known results in this or that scientific field, or formulate problems of a given scientific field in a system language (input, output, block, transfer function, state equations, etc.). This reinterpretation may become something of an intellectual excercise and appear a hobby (as someone once put it). Thus, it becomes a recreational activity for systems scientists, i.e. experts of methodologies. On the other hand, formulation of problems in systems language may become a 'virtuoso' exhibition for those who have a merely superficial knowledge of the systems approach and can thus hide gross mistakes. Obviously those who fall into the latter trap are experts in single specific sectors.

The setting-up of a model and its actual construction require a knowledge of the specific sector (economic, biological, etc.) and an equally deep knowledge of methodologies. In general, therefore, efficient model building can only come from collaboration between different experts. A large number of the reservations held against the construction of models and their relative lack of success (especially a few years ago) have come out of this lack of collaboration.

Considerations on the Relevance of the Interpretative Model

We have already pointed out that the systems approach focuses attention on behavioural models. This may lead to a concentration of research on identification methods of the black box which in practice are not usable unless we at least assume some a-priori hypotheses on the structure of such a box. This means that to some degree we have to make the box transparent. Here again, the important role of interpretation becomes apparent. That is why we must admit the obvious relevance that modelling owes to the interpretion of the mechanisms of interaction between input and output quantities, whatever the pros and cons of the dichotomy between interpretation and behaviour might be. Interpretation is the only non-aprioristic way to follow, as stated above, in order to reduce the system class on which identification can be performed.

The ever-increasing importance of the analysis of possible internal mechanisms is confirmed by the connection between such analysis and the success so far granted to modelling in different application sectors. It is sufficient to compare the validity of models shown in technological control system sectors and the doubts over that of mathematical models in the social sectors. These differencies are related to the diverse degree of development of quantitative theories in distinct sectors and to the diverse character of laws to which one can appeal (and that do not very often have the same sound experimental foundations as does physics).

In conclusion, mathematical modelling is generally an interdisciplinary task, in which systems scientists and problem specialists (e.g. economists, biologists) join together in a working group, each contributing his own particular skill. This does not mean that a problem specialist cannot become a good mathematical modeller in his field of interest. Yet, it is clear that good mathematical modelling skill requires a thorough mathematical background and a specific knowledge of systems science theories and techniques. Often enough, these are not prerequisites to becoming a good professional like, let us say, a biologist.

A problem specialist thus has usually a long way to go in

studying theories and techniques if he wants to become a mathematical modeller himself. The same applies the other way around. It is not a trivial matter for a system scientist to gain the insight into a particular problem necessary for him to build reliable mathematical models.

CONSIDERATIONS ON THE USE OF MATHEMATICAL MODELS

To evaluate the relevance of the use of mathematical models, one must distinguish, from the beginning, between the construction and the use of the model. Needless to say, building a model is fundamental to its use, but the latter phase does not necessarily follow the former. Nevertheless, the first phase, i.e. construction, has its own validity, especially when a widely accepted model does not yet exist. It really means identifying the relevant variables, quantifying them and making hypotheses on the functional links between them which will then undergo a quantitative check. The really significant step, which is basic for the progress of scientific knowledge, is the checking of the hypotheses and formulating new hypotheses when the previous ones show themselves unsatisfactory.

The use of mathematical models introduces a more complex consideration. While on one side a model may again be used to get a better knowledge of a phenomenon through the study of the model properties (stability, dynamic characteristics in the short and medium term, etc.), on the other it may be quite often used for control purposes. The use of a model for control may be carried out in different ways, ranging from purely qualitative techniques to the most sophisticated optimization. The use of one technique or another is not exclusively a matter of choice of the designer, but often depends on the model's reliability, on the techniques available and on a cost/benefit analysis of the different techniques. As far as model reliability is concerned, it is clearly of no value to make use of algorithms in order to get optimal controls when a model only qualitatively reproduces the dependencies of outputs from inputs. In this case it is sufficient to discover through simulation which control variables tend to produce the desired qualitative behaviour, as happens when one wants to change an undesired trend of an output variable. On the contrary, when a reliable model is available, the use of more sophisticated techniques resulting from control theory or mathematical programming is not only possible but sometimes advisable. Nevertheless, the use of optimization techniques has to be carefully examined. Going beyond linear programming or non-linear programming and dynamic optimization of small-dimensional systems, optimization algorithms show a high efficiency only in particular cases. That is why a large-scale system requires the substantial use of computing power or a major effort in producing ad-hoc computing algorithms. This effort must face both the situation in which we are acting and the advantages that an optimal solution would bring. In other words, when a new control policy is to be designed to take the place of a previous one, a sub-optimal control is sometimes satisfactory and quite often preferred by the user, especially if it is not drastically different from the previous one and produces a significant improvement in system performance. In addition, we have to consider whether or not the necessary effort in computation to move to an optimal policy from a sub-optimal one is then balanced by an adequate enhancement of benefits.

All this having been said, no doubt should arise as to the contribution, that has been made and will continue to be made by mathematical models to applications other than those merely aimed at understanding phenomena. Significant confirmation of this is evident from the analysis of the diffusion of mathematical models in the various applicative and disciplinary environments, some of which will be examined in more detail in the following chapters.

AREAS OF PREVALENT INTEREST IN SYSTEMS SCIENCE AND MODELLING

Both methodological and applied research need to be developed in order to offer adequate tools for the technological, economical and social problems that confront modern society. As far as methodological research is concerned, the identification sector deserves special attention. It is a fact that identification methodologies have become highly sophisticated, mainly for specific classes of problems. Nevertheless, we have to note that the development of further improvements without a serious re-examination of the basic hypotheses on which the methodologies stand, has been recently questioned. In many sectors, there has been shown a weak relationship between the sophistication of models and the quantity of available data. In effect, in more than one sector, better modelling work needs an intervention on data problems rather than any sophistication of methods. In other words, as far as methodologies are concerned, a sounder theory should be developed as to what we can expect from limited data, while efficient methods for extracting such information should be set up.

Large-scale systems are another area of particular research interest. One only has to think of socio-economic systems to understand the urgency of developing more appropriate methods and techniques of large-scale system identification and optimization. In effect, a model of a socio-economic system is a large-scale model consisting of at least 100 equations; it will typically contain 500 and occasionally more than 1000 [9].

Another problem that research should overcome is the widespread refusal by policy makers to use mathematical models. In our opinion this stems from the intermediary role that system modellers usually play. Appropriate tools (like interactive codes) allowing a direct and easy use of models by policy makers themselves, should be developed.

As far as fields of application are concerned, this book focuses particular attention on environmental and economic systems. Although somewhat arbitrary and certainly not exhaustive of the fields of potential and actual use of mathematical modelling, the choice was based on two main considerations. Firstly, environmental and economic systems present many a critical aspect in modern societies. Consequently, the need for deeper knowledge and control of such systems is quite generally recognized. Secondly, the successful applications reported so far demonstrate that these fields are fertile for the dissemination of systems science. Nevertheless, much work still remains to be done in order to fill the large gap existing between the mathematical techniques presently available and the demand for solution of actual problems.

NOTES

[1] A first process of abstraction which moves the centre of attention from interpretation to behaviour has already been seen in mathematical physics with the study of models designed to interpret different phenomena.

[2] The use of a computer is also essential in the construction phase of the model; this becomes clear from an analysis of the methods which can be adopted in construction.

[3] The problem is really more complex and should be examined according to the generality of the definition. It could, in fact, call for the availability of only one copy of the phenomenon on which an infinite number of experiments is to be effected; or else more copies of the phenomenon, possibly an infinite number, and on each of them experiments should be carried out.

REFERENCES

[1] A. Rosenbluetn and N. Wiener, 'The Role of Models in Science', *Philosophy of Science*, No. 12 (1945).

[2] W. J. Karplus, 'The Spectrum of Mathematical Modelling and Systems Simulation', in *Simulation of Systems*, ed. by L. Decker (North-Holland, Amsterdam, 1976).

[3] I. Gustavsson, 'Survey of Applications of Identification in Chemical and Physical Processes', Proceedings of the 3rd IFAC Symposium on Identification and Parameter Estimation (The Hague, 1973).
[4] G. A. Bekey and J. E. Beneken, 'Identification of Biological Systems: A Survey', *Automatica* **14** (1978), 41–7.
[5] V. C. Rideout, 'Simulation in Life Sciences', in *Simulation of Systems* (see above).
[6] K. S. Srikantan, 'Policy Application of Population Models: A Comparative Review', *Socio-Economic Planning Sciences* **11** (1977), 1–11.
[7] J. W. Forrester, 'Dynamics of Socio-Economic Systems', 6th World IFAC Congress (Boston, 1975).
[8] Y. Sawaragi and S. Ikeda, 'Identification Methods in Environmental Pollution Problems', Proceedings of the 4th IFAC Symposium on Identification and Parameter Estimation (Tbilissi, 1976).
[9] J. R. Klein, 'National Economic Management', 6th World IFAC Congress (Boston, 1975).

Dipartimento di Sistemi, Università della Calabria, Italy
Istituto di Automatica, Università di Roma, Italy

J. M. Gvishiani

Systems Sciences and Crucial Modern Problems

Throughout the entire history of human civilization science and technology have affected the life of society by contributing to the solution of certain problems, introducing other new ones, and generally changing the mode and style of people's lives. The fruits of scientific and technological advance may be utilized for both the benefit of mankind or its detriment, depending on a variety of factors, and largely on the specific socio-economic environment in different countries and in different historical periods. At present, the extent of the impact of science and technology on social development is enormous. Never before has humanity had available greater potential for the solution of the problems it faced. However, certain aspects of current scientific and technological progress are directly or indirectly conducive to the emergence and evolution of a host of extremely complex problems to be faced by the modern world.

Following the growth of scientific achievements, the scope of their application, and opportunities for their practical unilization, the role of science in the life of man and society has acquired an evergreater significance. Now, due to the aggravation of global problems, this issue has gained a new dimension, urging science not only to generate new approaches to the interpretation, of problems, but also to contribute actively the search for new ways of their solution. The future of science and of the entire human civilization depends to a large extent on the results of these efforts.

The deliberate address of science to global problems is often interpreted as an attempt to rehabilitate itself from the negative phenomena attributed to scientific and technological progress which have allegedly put mankind on the horns of a dilemma: either perish, following the path of further scientific and technological change generating irreversible negative results; or survive, by limiting technological development in the name of preventing the forthcoming 'global disaster'. What is more, the very fact of the emergence of global problems is at times viewed as an unpredicted, entirely negative result of human technological achievements, or the erroneous direction of the entire humanity's evolution ostensibly testifying to mankind's self-destruction.

We do not share the viewpoint that the global *problematique* results from a certain deviation of humanity from its historical course due to an excessive love for the values of industrialism which were misconceived and which became habitual. We are equally opposed to both the concepts of scientism and technocracy, laying hopes on science and technology as a magical means for the solution of our socio-economic problems, and of antiscientism which negates science, attributing to it an independent, self-contained role in modern society and blaming it for all its misfortunes. Science is a social institution related to the needs of social development and always reflecting the values and goals of a concrete society. It is the type of society that, in the final count, determines how the scientific achievements will be utilized, and to what extent they will contribute to solution of social problems.

Soviet researchers are studying global development problems along two major lines. First, they are conducting a comprehensive study of certain facets, forms and specific manifestations of the global *problematique* in individual areas and spheres of natural and societal development. The accumulation and generalization of empirical data extending knowledge of the laws of civilization development, is carried out primarily by the traditional scientific disciplines or in close cooperation with them.

Soviet scientists have attained rather promising results in the study of social aspects of environmental control and demographic problems. They are conducting research into a more efficient utilization of energy resources and the development of new technology capable of employing non-conventional energy sources, more rational utilization of natural resources, and so on.

It is these specific studies that provide a basis for the second direction of global research, i.e. a more profound cognition of general laws of evolution of the global system as an integral phenomenon.

The present-day level of research into complex current problems, including global ones, clearly indicates that the concept of 'science in general' or 'science as such', etc., is too abstract for discussion on any problem. On the other hand – and this is also common knowledge – the attempts to succeed in the study of, and search for, alternative solutions of such problems within the framework of individual scientific disciplines do not, and objectively cannot, lead to any progress. Hence, there arises an inevitable question concerning an effective interdisciplinary interaction, and the development of an up-to-date comprehensive research methodology. Even the relatively modest experience gained in the study of these processes shows that the idea of the interdisciplinary synthesis as a mechanical combination of isolated disciplines, tends to be fruitless. On the other hand, arguments about the 'interfaces' of sciences, 'bridges' between them, etc., are nothing more than the images which probably help express certain intuitive ideas verbally but do not promote actual knowledge of their essence. The question is much more complex, and its solution will probably require the use of the entire range of modern research tools: substantiated philosophical reasoning, specific socio-economic analysis, mathematical models, and the extensive accumulation of relevant empirical data. Only this way will lead to real progress towards a deeper understanding of the unity of modern science and its role in the solution of the crucial problems of today.

No-one now questions the need for a systems approach to the study of complex modern problems, tendencies in their development, and the goal-oriented control of these processes. However, the mere statement of this rather obvious fact by no means guarantees that such systems research will be practically realized: it may remain a good intention or, what is worse, happen to be just a facade concealing the application of obsolete traditional analytical techniques inadequate for the complexity of the system studied. The point is that systems analysis, systems approach and systems research are not mere fashionable words indicating that a researcher perceives the complexity of problems he faces. They represent a research methodology and, as with any other methodology, special knowledge and skills are required for its successful employment. Moreover, specific requirements for concrete methods of systems research into current problems are set by the nature of these problems and the peculiarities of the objectives facing the researcher. One of the scientific tools allowing one to implement the systems approach as applied to research into global problems is global modelling, the worldwide application of which has already yielded certain positive results. However, before proceeding to a discussion of global modelling and of the results obtained by Soviet researchers, it seems reasonable to outline briefly the systems approach and the principles of systems research.

During the last decade, a distinguishing characteristic of science and technology development was the study of complex systems of a different nature, the solution of multiple-factor and multiple-criteria problems combining the system's formalizable and hardly formalizable properties and characteristics, including the social 'human factors' in organic unity with economic and technical factors. All this made it necessary to adopt a really interdisciplinary approach, and to search for theoretical means such as notions, techniques, principles etc. and to employ computing technology in the study of such systems. Though the development of systems principles is effected both within the frameworks of separate disciplines (e.g. biology, sociology) and on interdisciplinary basis, on the whole the conceptual framework of systems research has a pronounced complex, integrative dimension stipulated by the unity of their cognitive and transforming functions. This unity, which is the major characteristic of systems research as an area of scientific activity, consists in a steady orientation of the latter towards cognition of not only and not so much of the phenomena proper, but rather of the methods of their practical mastering, and at the study of the essence of complex (including global socio-economic) problems along with the alternative ways of their solution.

The emerging tasks of the forecasting, formulation and comparison of alternatives, the validation of organization and management decisions, etc., are of a complex nature and require consideration of a variety of qualitative and quantitative factors affecting not only 'technological' but also 'social' functioning of the systems under study. Many problems of technological change management are related to medium- and long-term forecasting and planning, formulation of important national economic decisions, the study of regional and global development, and so on. Hence, they are quite systemic by content.

The nature of the problems have required the elaboration of a system of methodological tools both for relevant research and the decisions concerning all the evolving practical activities.

Methodology of this kind is represented in particular by the systems approach designed to generalize and express scientifically the key principles, notions and techniques of systems research as a whole. Bypassing the characteristics of the philosophical problems of systems research[1] we shall mention just two aspects that seem significant to the context of this article.

First, the systems approach is one of the general methodologies of research. This means that irrespective of its initial form, the systems approach is no longer oriented at any special science (important though it may be within the system of scientific knowledge) but at science as a whole, at the integration – naturally from a specific point of view – of the advances in the social, natural and engineering sciences, as well as at practical experience, primarily in the field of organization and management.

Following the tendency towards integration in present-day scientific knowledge, and in response to the requirements set by the specific features of the problems facing mankind, the development of systems research is directed toward a holistic and, accordingly, synthetic interpretation of the complex systems under study. It is this interpretation of a system through the synthesis of perceptions of its separate facets and elements that constitutes the objective and characteristic of the systems approach. Its ideal in this respect is a *systems synthesis* which embodies integration of special knowledge, and implements an actual interdisciplinary, synthetic perception of a system.

At the same time the general scientific nature of the systems approach establishes its very close relationship with philosophy. Without having a claim on the philosophical generality of its conclusions, the systems approach provides a link between the philosophical methodology of dialectic materialism and the methodology of special sciences. Being the reflection and the result of a multilevel structure of modern scientific cognition, this relationship of philosophy and special sciences, through general scientific notions and concepts, serves as one of the effective means of strengthening the union between dialectic materialistic philosophy and the social, natural and engineering sciences. It should be noted that the systems approach, as well as the general systems theory and systems analysis, can be considerably enriched by the further development of the philosophic principle of the systemic view, providing a basis for the philosophic ideological interpretation of the role and significance of systems research methodology.[2]

Second, the specific feature of systems methodology is its *direct orientation towards concrete practical problems.* The current global problems represent the most illustrative and significant (but far from unique) examples of such problems. Practical orientation is reflected in systems analysis[3] concerned primarily with the development of theoretical and methodological tools of research on ill-structured problems, systems engineering involving the human factor, and for the solution of these problems and management of these systems. Of special significance among the tools is the simulation technique. It should be noted that systems research implies both formalizable and non-formalizable model components without which it would be impossible to adequately reflect the complex relationships of the modelled problems and processes. Later in this article we shall consider this issue in greater detail.

Thus, systems research is designed to solve a triple task: to accumulate and express, in general scientific terms and concepts, the latest advances in social, natural and engineering sciences concerned with the systemic arrangement of real-life objects and ways of their cognition; to assimilate the principles and experience of evolution of dialectic materialistic philosophy, the resultant philosophic systemic principle and a number of related categories; and to apply the relevant conceptual framework and modelling tools to the solution of complex modern problems. Of special significance in this respect is the further development of applied methods realising the synthetic capabilities of the systems approach and providing for a systemic synthesis of knowledge about the processes under study, oriented towards theoretical and practical problems. One of the areas in urgent need of such interdisciplinary systems synthesis, is the study of global development problems with the help of systems modelling.

During the last decade, modelling of global development, or global modelling, has progressed from pilot experiments to an independent field of research performed by new and larger teams of specialists and large research institutions.

Construction of models reflecting the essential properties of a studied system and consistent with its objectives, is one of the principles. Such models, representing a certain holistic interpretation of the system and the interrelationships of its major subsystems and elements, are based on a certain concept which expresses the researchers' philosophy.

Application of the simulation technique was no accident, as it corresponds to its actual significance both in systems research and in modern science as a whole. We are safe in saying that today modelling, especially computer modelling, is the most valuable theoretical tool of applied systems research. This is conditioned by the unique features of the current stage of cognition and management of complex systems. The deep internal relationship between modelling and the systems approach is traced in the very way in which system consideration, as a system, representing the object under study, can be treated as the latter's model. On the other hand, treatment of a complex object's model as a system is often a rather effective method of its study. Combination of the systems approach and modelling is very fruitful as it enhances their interaction and mutual enrichment.

This necessitates a really systemic consideration of the system under study, involving the entire spectrum of essential factors of its development, as applied to a specific task. Otherwise, in cases where the basic premises and data are chosen in a one-sided, non-systemic and sometimes biassed fashion, the utilization of mathematical models – sophisticated though the hardware may be – will be just window-dressing of 'scientific objectivity' and in fact will not result in more than rather trivial explanations and often ineffective recommendations.

At the same time, if modelling is really science-based, it becomes possible to take account of the systemic and complex dynamics of the very processes under study, even the hardly formalizable aspects.

It is common knowledge that the successful solution of research problems is to a considerable degree determined by their clear, strict formulation. The mathematical modelling of socio-economic processes encounters a host of difficulties stemming from the properties of these systems. The most serious problem relates to the fact that we are unaware of the exact quantitative laws of development of the given process. We know empirically only some regularities in narrow fields of knowledge that can be correct under certain quite definite conditions. Hence, it is impossible to build a 'unified' model valid for a wide range of problems. Like the majority of systems analysis problems, each problem of global development study requires its own special model resulting from extensive interdisciplinary research staged by social, natural and engineering science experts.

Viewing global modelling as a sufficiently effective tool of global development studies, one must be aware of its limits and constraints. Two characteristic features should be noted with respect to this method. An important factor, considerably affecting the quality of results, is the level of model disaggregation, i.e. the extent of detail in process description. One would think that the more data we use, the more exact would be the conclusions. However, we have to take account of the fact that the larger the number of parameters with which we describe the behaviour of the global system (and hence the more complex the model becomes), the more we risk crossing the so-called 'interpretation threshold', i.e. losing the capability to perceive and estimate the results correctly.

On the other hand, the utilization of highly aggregated indicators is fraught with serious mistakes of another kind: the conscious or unconscious 'projection into the future' of those trends, structures, interrelations and ideas that seem customary at present. The 'paradigms of the present day' built into the highly aggregated models inevitably distort the picture of the future, since it is well known that evolution is far from being confined to purely quantitative changes. It implies a profound transformation of the very structure and interrelations between separate socio-economic processes, and it is these processes that play such a decisive role. Application of present views and values, as well as of available approaches to problems of the future will not only fail to provide any meaningful results but is fraught with serious errors.

This so-called 'historism of notions' is a major obstacle to developing formal theories that could be helpful in forecasting socio-political phenomena and processes within a reasonable time horizon. The inevitable subjective evaluative factor, much talked about by sociologists and political scientists, considerably affects the reliability of the data obtained.

It has already been pointed out that for a system of models to adequately reflect the tendencies of global system development, it must embody all significant processes, especially from the point of view of the future. Not all are prone to complete formalization. For example, among the partially formalizable processes are the primarily socio-political ones. At the same time, there are sufficiently good scientific tools for modelling economic processes. Attempts to formalize and model, in a long perspective, such non-formalized or non-formalizable processes, sometimes result in serious errors in conclusions and forecasts. An important criterion for choosing such 'model' processes is also the availability of reliable and homogeneous statistical data accumulated over a sufficiently long time and covering a wide spectrum of countries and regions.

The unfeasibility of applying the principles and methods effective in medium- and short-term economic forecasting, to longterm global development urges one to look for some other means and solutions.

One of the ways in overcoming the difficulties and modelling the phenomena and processes that are either non-formalizable at all or cannot be formalized with the present level of scientific knowledge is by using the scenario technique.

The variables which considerably affect the dynamics of global development but which are disposed towards strict formalization are brought together in a group of scenario variables. This allows one to formulate and study different options of global development, and evaluate the implications of various decisions. It is the deeper understanding of global development options rather than the development of allegedly reliable forecasts that is one of the major purposes of global modelling.

Such perception of research objectives was reflected in the actual implementation of 'Modelling of global development' project which has been conducted since 1977 by the All-Union Research Institute for Systems Studies (VNIISI) of the State Committee for Science and Technology of the USSR and the USSR Academy of Sciences. First of all, the project does not pursue the creation of any single global model, simulating the global system development and somehow or other defining the interdependence of its subsystems, but a *modelling system*. It contains a system of global development models, data bank, scenario formalization block, algorithm and programme block, result interpretation block, and, finally, an interaction system. This structure makes it possible to extend its research capabilities continually by changing, in accordance with the specific task, the number of models, increasing the data bank and improving the scenario tools.

One of the advantages of this modelling system is that it overcomes the problems that often emerge during interaction with the computer, i.e. either a 'computer fetishism' or development of 'computer complexes of inferiority'. This system, however, thanks to an elaborate interaction system, is suitable for any expert, even for one ignorant of either special machine language or translation principle.

The development of this quite new modelling system required a host of novel scientific and technological approaches. Some things have been successfully accomplished, others are coming. However, there are a number of problems we are so far unable to solve. Yet, a number of interesting problems can already be successfully solved with this modelling system today. The global development modelling system being worked on in VNIISI is based on the following principles:

(1) Man-machine approach – all phases of construction, study and utilization or the global development model are run in the man-computer interaction mode.

(2) Versatility – capability of the system to develop models for the study of dynamics of long-range global development; to store basic data related to the class of problems under study.

(3) Adaptivity – possibility of easily restructuring the original model, as well as research techniques consistent with newly formulated problems.

(4) Combination of qualitative and quantitative research techniques – taking advantage of formal modelling and informal capacities of man, employment of the maximum number of relevant factors, including the phenomena whose dynamics are not yet formalizable.

(5) Openness – ability of the system to change, improve through the increment and replacement of existing elements with new, improved blocks.

(6) Cumulativity – ability to accumulate knowledge as well as model, algorithm and statistical data.

(7) Specificity – ability to employ the general characteristics of the considered class of problems which the modelling system is designed for; then, on the basis of this specificity, to simplify the algorithms and programs employed and to increase their effectiveness.

(8) Simplicity of use – accessibility of the system to the researcher at all stages of modelling; possibility of arranging man-machine interaction so that the requests and answers are in a language close to natural, in terms and categories understandable for the researcher; possibility of working with the model and remaining ignorant of the system's structure; to obtain intermediate and final data in a form suitable for the interpretation of results.

At present the modelling system is used for the construction of a nine-region global development model, taking account of the mutual influence and dynamics of social, economic, ecological, demographic and other processes, as well as various options of global energy development. Today, we already have the first version of this model. It is instrumental in the running and study of alternative scenarios of global development related, in particular, to issues of peace and disarmament. It can also explore the switching of resources being spent for military purposes, to the needs of socio-economic development, different options of demographic development in separate regions and on the globe as a whole, etc. There are plans to study alternative scenarios of global development, given different levels of cooperation between various regions and nations. The degree and intensity of interregional cooperation considerably affect the solution of the most pressing global problems. Interesting results were obtained from the research staged with this modelling system on food supply of the growing population of the globe under different assumptions of population growth and agricultural production increase.

Conducting research in the field of systems modelling, one must be fully aware that a model, a group of models or even a modelling system does not provide ready answers, sound predictions or forecasts. As any tool, they are good only in skilful hands, and the quality of research they are used in is determined, in the final count, by the depth and substance of conceptual framework underlying any simulation experiment.

Perception of the long-range human development horizon is a distinguishing characteristic of the Marxist approach to the study of social development processes. It is in this broad context that we try to consider the managerial problems facing us today. The need to take account of a broad international context in planning the USSRs socio-economic development arises, in particular, due to the development of our international economic, scientific and technological ties – the very essence of our detente-oriented policy – and business-like, friendly contacts with all nations. Rather promising in this respect seems to be the application of the modelling system with a view to a comprehensive evaluation of longterm plans for the country's socio-economic development.

Recent years have witnessed more intensive work on scientific, technological and socio-economic forecasting in the USSR. Of greatest significance is the 'Comprehensive Programme of Scientific and Technical Progress for a Period of 20 Years' which is one of the most important pre-plan documents. Such a programme has been developed twice, on an experimental basis, jointly by the USSR Academy of Sciences, the USSR State Committee for Science and Technology, and leading research organizations of the country. At present the third stage of work on the Comprehensive Programme is being conducted, and the Programme has already been designed up to 2005. The Programme has been called to formulate concisely the most significant scientific, technological and socio-economic tasks for the long term whose solution will help raise the technological level of all branches of the national economy. It is quite obvious that in formulating the perspectives of the country's development we cannot help taking into account the general processes of global development and the global problems facing humanity – in other words the context of our progress. In this connection many aspects of global problems will be broadly reflected in the Programme being developed.

The ever-increasing complexity of socio-economic problems, which are acquiring a global nature and need active response and action on the part of science as an institution responsible for human cognition, makes it necessary to think more often about the global nature of the world science itself. The effective solution of global problems by science presupposes extensive international cooperation between scientists, as the traditional forms of scientific and technological cooperation, i.e. a simple exchange of experience and information, are inadequate since they leave the financing of research to national scientific institutions. The huge scale of pressing problems facing mankind requires a real 'brainstorming', the most extensive cooperation of scientists, both international and interdisciplinary. Indeed, the problems are so difficult and complex that they cannot be solved either by individual scientific disciplines or by separate, even the most industrially developed, nations.

The new trends in world science also show themselves in the development of new forms of research organization, simultaneously combining cooperation on interdisciplinary and international levels, which greatly enhances concentration of efforts and resources on the solution of the most pressing problems. These forms of cooperation involve international institutes engaged in the study of crucial problems of the present time.

For almost a decade there has existed the International Institute for Applied Systems Analysis (IIASA) in Vienna[4]. With regard to its research *problematique*, organization structure and research staff the Institute constitutes a new type of research organization based on international and interdisciplinary cooperation of researchers. The Institute staff consists of 80–100 scientists (20–25 from the USSR) from 17 countries of the world. IIASA collaborates with different national and international organizations actively contributing to its activities. Along with the elaboration of systems analysis methodology, IIASA carries out extensive applied research on such pressing global problems as the long-range development of world energy and the effective utilization of water resources. It analyses problems of health care, the organization and management of human settlements, and studies demographic problems.

The creation of IIASA signifies the origination of a new improved form of international collaboration of scientists from different countries. The Institute's activities not only enhance the further advance of science and technology but also contribute to detente, earning a reputation as the 'offspring of detente'.

In 1973 IIASA started research on an important programme entitled 'Energy systems'. It was headed by a prominent specialist in the energy field from the Federal Republic of Germany, Prof. W. Haefele. At different stages of the programme activities more than 140 experts from 20 countries including the USSR, contributed to it. The aim of the programme was the study of perspectives of the world energy development for the period 1980–2000.

The major conclusion of the IIASA research concerning the perspectives of global energy supply for the next fifty years is rather optimistic. It is supposed that during this period the world population will double and approach 8 billion. Even with extremely low rates of economic growth, rigid economy of energy and the introduction of energy-saving technologies, the total global energy demand will grow to three or four times the present level of consumption. Nevertheless, the interdisciplinary analysis made by an international group of IIASA researchers has shown that available resources and further technological advance can meet this considerably increased demand, though it would be ensured only through the maximum utilization of available resources, including those heavily polluting the environment and those with a high cost of extraction.

The aggravation of problems of a global character such as energy, raw materials and others, has made progressive public opinion in the world realize the interdependence of all the processes and phenomena now constituting an integral world process, the necessity of mutual understanding between peoples, and the need to for detente and cooperation.

In some sense the aggravation of global problems has affected mankind in the same way as stresses affect individuals. Until recently, physiologists considered stress a negative phenomenon and its effect on human organism destructive. Gradually, however, it has become evident that it is wrong to give a purely negative label to its effects. It has turned out that stress is quite necessary for the human organism as a means of adapting to new and unusual conditions. Probably, a similar effect will be exerted by the deteriorating global problems, which have served to illustrate practically for the first time the close interdependence between all the elements and processes of the global system, and to indicate that detente and cooperation are indispensable conditions not only for the solution of these specific problems but probably also for human life on earth.

NOTES

[1] These issues are treated, in particular, in the paper by J. M. Gvishiani 'Materialist Dialectics – Philosophical Basis for Systems Research', in a book *Systems Studies. Methodological problems*, Annual edition, 1979, Moscow, 1980.

[2] For a detailed description see: I. V. Blauberg, V. N. Sadovsky, and B. G. Judin, 'Filosovskii printsip sistemnosti i sistemny podhod' (*Voprosy filosofii*, 1978, No. 8). – Философский принцип системности и системный подход.

V. P. Kuzmin, 'Mesto sistemnogo podhoda v sovremennom nauchnom poznanii i marksistskoi metodologii' (*Voprosy filosofii*, 1980, Nos. 1, 2).–Место системного подхода в современном научном познании и марксистской методологии.

[3] The term 'systems analysis' is not at all indicative of the greater share of analytical techniques at the expense of synthetic ones, and the word 'analysis' should not mislead anyone. This is rather a reflection of the long predominantly analytical orientation in the evolution of scientific cognition that has almost turned the notion of 'analysis' into a synonym of scientific research. Leaving aside the terminological arguments, we shall note that the combination of terms 'systems' and 'analysis', as applied to the characteristics of the given line of research, actually contains at least a certain contradiction, if not a paradox.

[4] The final document of the first phase of the IIASA programme was published in the form of a two-volume book entitled: *Energy in the World with Limited Resources*: *Global Energy Analysis*, highlighting the activities for the period 1973 to 1979. The book was published in 1981 in the USA. It was translated into Russian and will soon be in print in the USSR.

Institute of Systems Studies, Moscow, U.S.S.R.

G. Guardabassi

Interconnected Systems Analysis

INTRODUCTION

This article deals with interconnected dynamical systems with special attention being paid to their analysis. This means that a model consisting of interacting submodels is assumed to be available and the typical objective is to reduce the systems analysis into a sequential procedure, each step of which calls, in turn, for the analysis of either a part only of the original system, namely one of its subsystems, or a less detailed description of the overall system, namely a reduced-order system in which a suitable (possibly simplified) description of the original interconnecting structure is also incorporated.

The first property we shall be interested in will be the existence and uniqueness of the system output for any admissible input and initial state (well-posedness). This is, in fact, the least we can ask in order that a mathematical model stands as a proper description of the dynamics of a physical system. More precisely, we shall see that by no means every interconnected system made up of well-posed subsystems is well posed. Hence, some two-step criterion of the aforementioned kind will certainly be of interest.

Next, we shall look at the problem of actually computing the output, over a given interval of time, for any admissible input and initial state (simulation). More specifically, we shall consider the following situation. Assuming that the available computing facility is not powerful enough to handle the overall interconnected system at once (actually assuming that it cannot handle much more than a single subsystem at a time), is it possible to devise an algorithm by which the overall simulation problem can definitely be solved in a finite number of steps? In other words, the question of interest here is under what conditions can the overall simulation problem be solved 'subsystemwise'.

One of the central issues in interconnected systems analysis is, of course, the stability of the overall system. Typical results in this area are in fact a variety of criteria ensuring, in different settings, an appropriate kind of stability for the overall system, based on stability properties of the component subsystems combined with suitable features of the mutual interaction existing among them. In other words, most of these results are different formal phrasings of the intuitive idea that the stability of the overall systems is preserved if the isolated subsystems are stable and the mutual interaction among them is, in a proper sense, sufficiently weak. Decreasing the computational effort needed to test for stability and gaining a better understanding of some structural factors affecting the system stability are the main advantages associated with the aforementioned kind of two-step procedures. The price for this is a certain conservatism of many criteria.

The last aspect we shall consider in dealing with large-scale interconnected systems is the possibility of alleviating some of the difficulties involved in the system analysis by neglecting first some small or secondary effect (approximate reformulation) and then improving this approximation by reintroducing, in a sense, a significant part of the neglected factors through suitably adapted and computationally cheap asymptotic expansion techniques. The standard situation suited for perturbation analysis is one in which the system under consideration is affected by a small parameter. Setting such a parameter equal to zero simplifies the analysis in (at least) two different ways: either tearing some crucial links among the subsystems (decoupling) or reducing the order of the system. The two situations are usually referred to as regular or singular perturbation, respectively. The system behaviour with the small parameter set equal to zero may often (yet cautiously) be taken as a zeroth order approximation to the behaviour of the original system. By asymptotic expansion techniques it may then be possible to consider higher order approximations, thus improving the accuracy of the results by a moderate computational effort.

WELL-POSEDNESS AND SIMULATION

According to the plan outlined in the introduction, this article deals with interconnected dynamical systems; namely, with systems the state vector of which is partitioned into a number of subvectors so as to produce, in a straightforward way, a corresponding decomposition into dynamical subsystems of the considered system.

To make things as simple as possible, let us discuss well-posedness of interconnected systems in the context of discrete-time dynamical systems, i.e. of systems of the form

$$S: \begin{cases} x(t+1) = f(x(t), u(t), t) \\ y(t) \quad\;\; = g(x(t), u(t), t), \end{cases}$$

where the time t ranges over the zet Z of integer numbers, while $u(t) \in R^m$, $x(t) \in R^n$ and $y(t) \in R^p$ are the input, state and output vectors, respectively. If f and g are given functions mapping $R^n \times R^m \times Z$ into R^n and R^p, respectively, then the system S is apparently *well-posed* in the sense that, for any interval $T \triangleq \{\bar{t}, \bar{t}+1, \ldots, \bar{t}+L\} \subseteq Z$, any initial state $x(\bar{t}) \in R^n$ and any input function u defined over T, the state and the output motions x and y of S over T exist and are unique. It is, however, easy to see that even if we connect only well-posed systems the resulting overall system may well be not well-posed. As an example consider

$$S_1: \begin{cases} x_1(t+1) = u_1(t) \\ y_1(t) \quad\;\; = x_1(t) + 2\,\mathrm{sign}\,(u_1(t)), \end{cases}$$

$$S_2: \begin{cases} x_2(t+1) = u_2(t) \\ y_2(t) \quad\;\; = x_2(t) + \alpha u_2(t), \end{cases}$$

where all variables are scalar while sign (u) is by definition equal to 1, 0 or -1 according to whether u is positive, zero or negative respectively. Assume that S_1 and S_2 are connected by

$$I: \begin{cases} u_1(t) = y_2(t) \\ u_2(t) = y_1(t), \end{cases}$$

so that the composite system $S = (S_1, S_2; I)$ is free (no input) and let $\bar{t} = 0$, $L = 5$, $x_1(0) = 0$, $x_2(0) = 1$. Then it is easy to check that

(i) if $\alpha = 1$, the motion of S is not unique; more specifically, there are two possible (output) motions of S:

Motion No. 1:	$y_1(0) = 2$	$y_2(0) = 3$
	$y_1(1) = 5$	$y_2(1) = 7$
	$y_1(2) = 9$	$y_2(2) = 14$
	$y_1(3) = 16$	$y_2(3) = 25$
	$y_1(4) = 27$	$y_2(4) = 43$
	$y_1(5) = 45$	$y_2(5) = 72$
Motion No. 2:	$y_1(0) = -2$	$y_2(0) = -1$
	$y_1(1) = -3$	$y_2(1) = -5$
	$y_1(2) = -7$	$y_2(2) = -10$
	$y_1(3) = -12$	$y_2(3) = -19$
	$y_1(4) = -21$	$y_2(4) = -33$
	$y_1(5) = -35$	$y_2(5) = -56$

(ii) if $\alpha = -1$, there does not exist any motion of S,

(iii) if $\alpha = 0.5$, the motion of S exists and is unique; correspondingly, the output motion of S (over T) is as follows:

$$\begin{array}{ll} y_1(0) = 2 & y_2(0) = 2 \\ y_1(1) = 4 & y_2(1) = 4 \\ y_1(2) = 6 & y_2(2) = 7 \\ y_1(3) = 9 & y_2(3) = 10.5 \\ y_1(4) = 12.5 & y_2(4) = 15.25 \\ y_1(5) = 17.25 & y_2(5) = 21.125 \end{array}$$

Given any large-scale interconnected dynamical system, a detailed analysis of the kind performed on the above elementary example would in general be extremely cumbersome if not plainly impossible. We are therefore interested in any criterion enabling us to take some advantage of the system structure, i.e. telling us something about the well-posedness of the overall system in terms of data obtained by analysing first one-by-one the component subsystems and then gathering this scattered information into a suitably simplified general view of the overall system. One such criterion has, in fact, been proposed by Vidyasagar [1]. In order to introduce it in the simplest possible way, let us consider a discrete-time interconnected dynamical system $S = (S_1, S_2, \ldots, S_N; I)$ of the form:

$$S_k: \begin{cases} x_k(t+1) = f_k(x_k(t), u_k(t), t) \\ y_k(t) \quad\; = g_k(x_k(t), u_k(t), t), \end{cases}$$

$$I: \begin{cases} u_k(t) = \gamma_k(y_0(t), u(t), t), \quad k = 1, 2, \ldots, N, \\ y(t) \;= \sigma(y_0(t), u(t), t) \end{cases}$$

where, for any finite integers N and L, every $k \in \{1, 2, \ldots, N\}$ and every $t \in T \triangleq \{\bar{t}, \bar{t}+1, \ldots, \bar{t}+L\}$, $x_k(t)$, $u_k(t)$ and $y_k(t)$ are the state n_k-vector, the input m_k-vector and the output p_k-vector of the kth subsystem S_k, at time t; $x \triangleq |\, x_1' \, x_2', \ldots, x_N' \,|'$

is the state n-vector of the overall system S, $n \triangleq \Sigma_{k=1}^{N} n_k$; $y_0 \triangleq |y_1' \, y_2', \ldots, y_N'|'$ is the internal output p_0-vector, $p_0 \triangleq \Sigma_{k=1}^{N} p_k$'; u and y are the input m-vector and the output p-vector, respectively, of the overall system S.

In order to visualize S and, in particular, to enhance some important features of the pattern according to which the information flows through its subsystems, it is convenient to associate with S a time-varying digraph $G(t)$ called the *system graph of S at time t*. We may construct $G(t)$ in the following way. First we associate each variable u_k and y_k, $k = 1, 2, \ldots, N$, with as many vertices, called vertex u_k and vertex y_k, respectively. Next, we draw an arc from vertex u_k to vertex y_k, $k = 1, 2, \ldots, N$; these arcs will henceforth be called subsystem arcs. Furthermore, we draw an arc from vertex y_j to vertex u_k if and only if u_k does in fact depend upon y_j, i.e. if and only if $\varphi_{kj}(\cdot, y_0, u, t)$ is nonconstant for some $(y_0, u) \in R^{p_0} \times R^m$, where, for any $z \in R^{p_j}$,

$$\varphi_{kj}(z, y_0, u, t) \triangleq \gamma_k(|y_1' \, y_2', \ldots, y_{j-1}' z' y_{j+1}', \ldots, y_N'|', u, t);$$

these arcs will henceforth be called interconnecting arcs. Of course, two or more variables known to be identical at t (because of I) can be associated with the same vertex, which will be arbitrarily given the name of one of them.

The *reduced system graph of S at time t*, denoted $G_R(t)$, is obtained from $G(t)$ by removing each subsystem arc corresponding to a purely dynamic subsystem, i.e. to a subsystem S_k in which $g_k(x_k, \cdot, t)$ is constant for all x_k.

Since f_k and g_k, $k = 1, 2, \ldots, N$ are given functions defined over $R^{n_k} \times R^{m_k} \times T$, each subsystem S_k is obviously well-posed, in the sense that x_k and y_k over T exist and are unique, for every given initial condition $x_k(\bar{t})$ and every input function u_k over T. The question we are interested in is whether or not the overall system S is well-posed; in other words, we are looking for global well-posedness conditions somehow reflecting the composite nature of S.

In order to introduce one such condition, we need some further definitions. An interconnected system $\tilde{S} = (\tilde{S}_1, \tilde{S}_2, \ldots, \tilde{S}_N; \tilde{I})$ is to be a *structure-invariant perturbation* of S over T if:

(i) $\tilde{m} = m$; $\tilde{p} = p$; $\tilde{m}_k = m_k$; $\tilde{n}_k = n_k$, $\tilde{p}_k = p_k$, $k = 1, 2, \ldots, N$;

(ii) the reduced system graph of $\tilde{S}$ coincides with the reduced system graph of S, over T.

S is said to be *structurally well-posed* over T if all the structure-invariant perturbations of S over T are well-posed. Of course, structural well-posedness implies well-posedness but, contrary to well-posedness, depends on elementary properties of the (reduced) system graph only. In fact, the following result can be proved.

THEOREM 1. The interconnected system S is structurally well-posed over T if and only if $G_R(t)$ is acyclic (i.e. does not contain any closed path), for all $t \in T$. □

The role of structural well-posedness in interconnected systems analysis can be further clarified by looking at the problem of simulating S by a digital computer. Specifically, we assume that all functions f_k, g_k, γ_k, $k = 1, 2, \ldots, N$, and σ are computable by the available computing machine; in other words, we assume that an algorithm is available for each one of these functions, whereby they can be evaluated (up to any desired degree of accuracy) in any point of their domain by executing a finite number of computer instructions.

The system S is said then to be *simulable* over T if, for any initial state and any input function defined over T, the set of possible output motions of S over T is computable that is can be evaluated (up to any desired degree of accuracy) by executing a finite number of computer instructions. According to the adopted definition, every subsystem S_k of S is apparently not only well-posed but also simulable, yet we have already seen an elementary example ($N = 2$, $L = 5$) which, e.g. for $\alpha = 1$, is simulable though not well-posed. Since there cannot exist an algorithm to solve *any* set of nonlinear algebraic equations (even in case we know *a priori* that the set of all possible solutions is finite and nonempty), there must exist nonsimulable well-posed interconnected systems as well. Thus, the conclusion must be drawn that well-posedness and simulability are indeed somewhat related yet different concepts.

Suppose now that the simulation of S has to be carried on by making use of a computing facility whose central unit can efficiently handle the interaction I plus any subsystem S_k of S, but is not capable of efficiently handling (simulating) the system S as a whole, due to its exceedingly large order n. We are interested, then, in the following problem (subsystemwise simulation problem).

Given T, namely $\bar{t}$ and L, $\bar{x} \triangleq x(\bar{t})$ and u over T, compute y over (any right subinterval of) T by handling no more than a single subsystem at a time.

We say that S is *subsystemwise simulable* over T if the subsystemwise simulation problem can be solved by executing a finite number of computer instructions, i.e. if an algorithm for the subsystemwise simulation problem does exist. Finally, S is said to be *structurally subsystemwise simulable* over T if all structure-invariant perturbations of S are subsystemwise simulable over T. In [2], various algorithms are given for the subsystemwise simulation of structurally well-posed interconnected systems. Their relevance is qualified by the following result.

THEOREM 2. The interconnected system S is structurally subsystemwise simulable over T if and only if it is structurally well-posed over T. □

The most useful consequence of Theorem 1 and Theorem 2 is that, if $G(t)$ is acyclic for all $t \in T$, then S is well-posed and can be simulated subsystemwise. On the other hand, if this is not the case, it is fairly easy to see how the modify S slightly so as to make it structurally well-posed and hence subsystemwise simulable.

STABILITY

There are basically two approaches to stability analysis of interconnected dynamical systems: the Lyapunov approach and the input-output approach. Both of them have attracted the interest of several investigators working in engineering, mathematics, economics and biology. Two books [3, 4] are representative of the large amount of work carried on along the aforementioned streams. Despite certain similarities and numerous overlappings among the myriad of particular results obtained in the field during the last fifteen years, it is certainly impossible to give in a few pages a proper account of the state of the art in this so widely and carefully explored area. Therefore, we shall present only two typical and significant results wherefrom many others can be obtained either as special cases, or by slight modifications or even as natural extensions and generalization. In doing this, some familiarity with the standard Lyapunov or input-output stability concepts is assumed on the part of the reader. Following most of the existing literature, the presentation will refer to continuous-time models; similar results for the discrete-time case can be found, for instance, in [3].

The Lyapunov Approach

Consider a (free) continuous-time interconnected dynamical system $S = (S_1, S_2, \ldots, S_N; I)$ of the form

$$S_k: \begin{cases} \dot{x}_k(t) = f_k(x_k(t), t) + z_k(t) \\ y_k(t) = g_k(x_k(t), t) \end{cases}$$

$$I: \qquad z_k(t) = \gamma_k(y_0(t), t), \qquad k = 1, 2, \dots, N,$$

where, for any $t \in T = R$, $x_k(t) \in R^{n_k}$ and $y_k(t) \in R^{p_k}$ are the state and the output vectors, respectively, of the subsystem S_k of S, while $z_k(t) \in R^{n_k}$ represents the interaction. Furthermore, let

$$y_0 \triangleq | y_1' y_2', \dots, y_N' | \in R^{p_0} \; p_0 = \sum_{k=1}^{N} p_k.$$

We assume that S is well-posed. We also assume that the origin of the state space is an equilibrium state for each *isolated subsystem* S_k (with $z_k = 0$) as well as for the interconnected system S. In other words, we assume that, for every $k = 1, 2, \dots, N$,

$$f_k(0, \cdot) = 0,$$
$$\gamma_k(g_0(0, \cdot), \cdot) = 0,$$

where $g_0 \triangleq | g_1' g_2', \dots, g_N' |'$. As is well known, if S admits an equilibrium state $\bar{x} \neq 0$, it is always possible to carry S into the setting above by simply adopting $x(t) - \bar{x}$ as the new state variable.

Recall that the equilibrium $x = 0$ of S is *uniformly asymptotically stable* (in the sense of Lyapunov) if, denoting by $| \cdot |$, the Euclidean norm, for every $\epsilon > 0$ and $\bar{t} \in T$, there exists $\delta(\epsilon) > 0$, such that $|x(\bar{t})| < \delta(\epsilon)$ implies $|x(t)| < \epsilon$ for all $t \geqslant \bar{t}$ and $x(t) \to 0$ as $t \to \infty$. If $x = 0$ is a uniformly asymptotically stable equilibrium, its *domain of attraction* at $\bar{t}$ is the set of all $\bar{x} \in R^n$ such that $x(\bar{t}) = \bar{x}$ implies $x(t) \to 0$ as $t \to \infty$. The system S (any isolated subsystem S_k, $z_k = 0$) is said to be *globally stable* if the equilibrium state $x = 0$ ($x_k = 0$) is uniformly asymptotically stable and its domain of attraction coincides with the entire state space of $S(S_k)$.

A real-valued function $v(x, t)$ is said to be *positive-definite* over $R^n \times T$ if there is a nondecreasing real-valued function $\varphi(|x|)$ such that $\varphi(0) = 0$ and $0 < \varphi(|x|) \leqslant v(x, t)$, for all t and $x \neq 0$. If $\varphi(|x|) \to \infty$ as $|x| \to \infty$, then the positive-definite function $v(x, t)$ is *radially unbounded*. Finally, $v(x, t)$ is said to be *decrescent* if there is a nondecreasing real-valued function $\psi(|x|)$ such that $\psi(0) = 0$ and $v(x, t) \leqslant \psi(|x|)$, for all t and $x \neq 0$.

THEOREM 3. The interconnected system S is globally stable if

(i) for each isolated subsystem S_k ($z_k = 0$) there are a positive scalar α_k, a positive-definite function $w_k(x_k)$ and a decrescent radially unbounded positive-definite function $v_k(x_k, t)$, with continuous partial derivatives on $R^{n_k} \times T$, such that

$$\frac{\partial v_k}{\partial t}(x_k, t) + \frac{\partial v_k}{\partial x_k}(x_k, t) f_k(x_k, t) \leqslant -\alpha_k w_k^2(x_k)$$

and

$$\left| \frac{\partial v_k}{\partial x_k}(x_k, t) \right| \leqslant w_k(x_k)$$

$$\text{for all} \quad (x_k, t) \in R^{n_k} \times T;$$

(ii) for each $k = 1, 2, \ldots, N$, there are N nonnegative scalars β_{kj}, $j = 1, 2, \ldots, N$, such that

$$|\gamma_k(g_0(x, t), t)| \leqslant \sum_{j=1}^{N} \beta_{kj} w_j(x_j)$$

$$\text{for all} \quad (x, t) \in R^n \times T;$$

(iii) the $N \times N$ matrix A, the elements of which are given by

$$a_{kj} = \begin{cases} \alpha_k - \beta_{kk}, & \text{if} \quad k = j, \\ -\beta_{kj}, & \text{otherwise,} \end{cases}$$

is a Metzler matrix, i.e.

$$D_k \triangleq \text{de}\begin{pmatrix} a_{11} & \cdots & a_{1k} \\ \vdots & & \vdots \\ a_{k1} & \cdots & a_{kk} \end{pmatrix} > 0, \qquad \text{for all } k = 1, 2, \ldots, N. \qquad \square$$

A proof of Theorem 3 can be found in [5]. As for its basic interpretation and use, a few remarks are in order. First of all, note that condition (i) implies global stability for the isolated subsystem S_k; v_k is in fact a Lyapunov function for S_k. Condition (ii) involves the interconnecting structure only. A qualitative trade-off effect between subsystems and interconnecting structure is clearly expressed by the different roles played by the functions w_k in (i) and (ii). In any case, it is fairly apparent that, thanks to Theorem 3, the stability analysis of S can be accomplished in terms of qualitative properties of the lower order isolated subsystems S_k, expressed as individual Lyapunov functions, and of corresponding properties (bounds) of the interconnecting structure. Condition (i) ensures that, when completely decoupled, the subsystems are globally stable. Conditions (ii) and (iii) ensure that the coupling among the subsystems produced by the interconnecting structure is weak enough as to preserve global stability for S. Due to the very nature of this analysis, the condition provided by Theorem 3 tends to be conservative. On the other hand, it is well known that in some cases strong coupling is expedient to ensure global stability of the overall system. To deal with such a situation effectively (by Theorem 3 or by any similar stability condition) the only way is to incorporate each pair of strongly coupled subsystems of S into a single aggregated subsystem, thus turning in fact to a coarser decomposition of S. Of course, the price to pay, in this case, is an increased difficulty in obtaining the appropriate Lyapunov functions for the higher dimensional aggregated subsystems. In other words, the finer the decomposition of S, the easier is obtaining Lyapunov functions (if any) for the isolated subsystems but the more conservative is the condition then expressed by Theorem 3.

The Input-Output Approach

First of all we briefly recall a few notions needed to describe the general setting in which this approach has been developed. The basic feature, here, is that systems are viewed as operators from the input to the output function space. For more detail see [3] and [6].

Denote by L_p the space of real valued functions f over $R^+ \triangleq (0, \infty)$ with norm

$$\|f\|_p \triangleq \left(\int_0^\infty |f(t)|^p \,\mathrm{d}t\right)^{1/p} < \infty, \qquad 1 \leqslant p < \infty.$$

For any $f \in L_p$ and any $\tau \in R^+$, the truncated function f_T is defined by $f_T(t) = f(t)$ for $t \leqslant \tau$ and $f_T(t) = 0$ for $\tau < t < \infty$. L_{pe} is the extended space of all functions f such that $\|f_T\|_p < \infty$, for all $\tau \in R^+$. An operator A on L_{pe} is *nonanticipative* if $(Af_T)_T = (Af)_T$, for all $f \in L_{pe}$ and all $\tau \in R^+$, A is *bounded* if, for each $r_1 \in R^+$, there is $r_2 \in R^+$ such that $\|(Af)_T\|_p < r_2$ whenever $\|f_T\|_p < r_1$ and $\tau \in R^+$. The gain of A is

$$\mu \triangleq \inf\{\alpha\colon \|(Af)_T\|_p \leqslant \alpha \|f_T\|_p, \text{ for all } f \in L_{pe} \text{ and } \tau \in R^+\}.$$

If μ is finite then A is bounded. Any dynamical system described by an operator A on L_{pe} is said to be *input-output stable* if A is bounded. Note that if the system is either memoryless or linear and time-invariant, its gain can be readily computed from the input-out graph or from the Nyquist plot, respectively [6]. The extension of the definitions above to encompass the case of multivariable systems, i.e. of operators acting between vector-valued function spaces comes quite natural and is omitted as obvious.

Within this framework, we now consider a continuous-time interconnected dynamical system $S = (S_1, S_2, \ldots, S_N; I)$ described by

$$S_k: \quad y_k = G_k u_k + g_k$$

$$I: \quad \begin{cases} u_k = H_k u + \sum_{j=1}^{N} H_{kj} y_j + h_k, & k = 1, 2, \ldots, N, \\ y = Qu + \sum_{j=1}^{N} Q_j y_j + q \end{cases}$$

where u, u_k, y_k, g_k, h_k, y and q are vector-valued functions defined on R^+ while G_k, H_{kj}, Q_j and Q are nonanticipative operators on L_{pe}, for all $k, j = 1, 2, \ldots, N$.

We assume that S is well-posed; i.e. that, for any input u in the appropriate L_{pe} space, the internal input $u_0 \triangleq |u_1' u_2' \ldots u_N'|'$ and output $y_0 \triangleq |y_1' y_2' \ldots y_N'|'$ exist, are unique and depend causally on u so that the same is true for y and S can be viewed as a nonanticipative operator F; i.e. $y = Fu$. Analysing the input-output stability of S amounts then to finding conditions under which F is bounded. Again, this can be done in terms of properties of the isolated subsystems S_k and the interconnecting structure I. One such result, proved in [7], is as follows.

Suppose that all the operators appearing in S have finite gains, g_k, h_k, $k = 1, 2, \ldots, N$, and q belong to L_{pe} spaces of appropriate dimensions and let A be a $N \times N$ real matrix, the elements of which are given by

$$a_{kj} = \begin{cases} 1 - \mu_k \gamma_{kk}, & \text{if} \quad k = j, \\ -\mu_k \gamma_{kj}, & \text{otherwise,} \end{cases}$$

where μ_k is the gain of G_k and γ_{kj} is the gain of H_{kj}, $k, j = 1, 2, \ldots, N$.

THEOREM 4. The interconnected system S is input-output stable if A is a Metzler matrix (see Condition (iii) of Theorem 3). □

In dealing with the Lyapunov approach we have already observed that the price to pay for the advantage of decomposing the analysis of the overall system S into fairly independent steps, roughly associated with the isolated subsystems and their interconnection is a tendency of these results to be conservative. The same is true for Theorem 4. However, there is some evidence that input-output methods may occasionally be less conservative and easier to apply than Lyapunov methods.

SYSTEM APPROXIMATION BY PERTURBATION TECHNIQUES

Since the computational effort needed to solve problems involving dynamical systems generally grows at a greater than linear rate as the order of the system increases, there is a definite interest in finding approximate reformulations of the dynamical systems under consideration so as to produce the decomposition of the original problem into a sequence of subproblems involving each either a subsystem or a reduced order model of the dynamical system at hand.

Perturbation techniques are perhaps the best known, long established, general and powerful methodology available to produce in a very systematic way the decomposition above. They can be applied when the problem to solve involves a large-scale dynamical system affected by a 'small parameter' ϵ, the annihilation of which splits the problem into a set of 'at most sequentially interconnected' subproblems. It is then typical of perturbation analysis to provide means by which the small parameter effect is partially reintroduced without substantially affecting the sequential decomposition induced by setting the small parameter equal to zero. In other words, it is assumed that a power series expansion of the problem solution with respect to ϵ exists, then the zeroth order approximation obtained by setting ϵ equal to zero is improved by computing a finite number of higher order terms. The interesting point is that the computational effort needed to compute each additional term in the series above being roughly comparable with (usually smaller than) the one needed to compute the zeroth order approximation may be compared very favorably with the one called for by the exact solution of the original problem.

The approximation-decomposition mechanism induced by perturbation techniques cannot be appreciated in depth if no specific reference is made to the particular features of a well identified class of problems. In this respect, the richest framework is perhaps provided by estimation, prediction and control (or decision) problems. However, in order to make the presentation as simple as possible, only the problem of effectively computing the system output for a given input and initial state (simulation) will be considered here together with some basic stability issue.

There are two essentially different kinds of perturbation techniques. The first one applies when setting the small parameter ϵ equal to zero does not change the order of the system (regular perturbation); the second and more important one applies, on the contrary, to situations where setting ϵ equal to zero leads to a reduced order system (singular perturbation).

Regular Perturbation

The typical situation suited to be dealt with by regular perturbation techniques occurs when a small parameter ϵ can be used to describe a kind of 'weak coupling' among the subsystems.

Consider a time-invariant continuous-time interconnected dynamical system $S_\epsilon = (S_1, S_2, \ldots, S_N; I_\epsilon)$ of the form

$$S_k: \begin{cases} \dot{x}_k(t, \epsilon) = f_k(x_k(t, \epsilon), u_k(t, \epsilon)) \\ y_k(t, \epsilon) = g_k(x_k(t, \epsilon), u_k(t, \epsilon)) \end{cases}$$

$$I_\epsilon: \begin{cases} u_k(t, \epsilon) = \gamma_k(y_0(t, \epsilon), u(t), \epsilon), & k = 1, 2, \ldots, N, \\ y(t, \epsilon) = \sigma(y_0(t, \epsilon), u(t)) \end{cases}$$

where $y_0 \triangleq |y_1' y_2' \ldots y_N'|'$. Note that a (time-invariant) system graph G_ϵ can easily be associated with S_ϵ in exactly the same way as on page 37 and assume that the small parameter ϵ describes a weak cyclic interaction among the subsystems; in other words, assume that G_0 is acyclic. Furthermore, assume that S_ϵ is well-posed and sufficiently smooth (all functions used to describe it are $(1 + M)$ times continuously differentiable, for some $M \geqslant 0$).

We are interested in the following simulation problem. For any given input u, defined over a finite interval $T = [0, L] \subseteq R$, and initial state $\bar{x} \triangleq |\bar{x}_1' \bar{x}_2' \ldots \bar{x}_N'|'$, compute a Mth order approximation of the output y over T; i.e. a function $\tilde{y}(\cdot, \epsilon)$ such that $|y(t, \epsilon) - \tilde{y}(t, \epsilon)|/\epsilon^M$ goes to zero as ϵ goes to zero, for all $t \in T$.

A zeroth order approximation is readily shown to be provided by S_0. In what follows, we consider the case $M = 1$ (the extension to higher order approximations is conceptually straightforward). Letting

$$x_k^0(t) \triangleq x_k(t, 0), \quad u_k^0(t) \triangleq u_k(t, 0), \quad y_k^0(t) \triangleq y_k(t, 0),$$

$$\delta_k^x(t) \triangleq \frac{\partial x_k}{\partial \epsilon}(t, 0), \quad \delta_k^u(t) \triangleq \frac{\partial u_k}{\partial \epsilon}(t, 0), \quad \delta_k^y(t) \triangleq \frac{\partial y_k}{\partial \epsilon}(t, 0),$$

$$f_{kx} \triangleq \frac{\partial f_k}{\partial x_k}, \quad g_{kx} \triangleq \frac{\partial g_k}{\partial x_k}, \quad f_{ku} \triangleq \frac{\partial f_k}{\partial u_k}, \quad g_{ku} \triangleq \frac{\partial g_k}{\partial u_k},$$

$$\gamma_k^0(y_0, u) \triangleq \gamma_k(y_0, u, 0); \quad \gamma_{ky}^0(g_0, u) \triangleq \frac{\partial \gamma_k}{\partial y_0}(y_0, u, 0),$$

$$\gamma_{k\epsilon}^0(y_0, u) \triangleq \frac{\partial \gamma_k}{\partial \epsilon}(y_0, u, 0),$$

elementary calculus (since δ^x, δ^u and δ^y are often referred to as the sensitivities of x, u and y with respect to ϵ, this is also called sensitivity analysis) immediately leads to considering the interconnected system $\tilde{S}_\epsilon = (S_1^0, S_2^0, \ldots, S_N^0; S_1', S_2', \ldots, S_N'; \tilde{I}_\epsilon)$, where

$$S_k^0: \begin{cases} \dot{x}_k^0(t) = f_k(x_k^0(t), u_k^0(t)), & x_k^0(0) = \bar{x}_k, \\ y_k^0(t) = g_k(x_k^0(t), u_k^0(t)) \end{cases}$$

$$S_k': \begin{cases} \dot{\delta}_k^x(t) = f_{kx}(w_k^0(t), u_k^0(t))\,\delta_k^x(t) + \\ \qquad + f_{ku}(w_k^0(t), u_k^0(t))\,\delta_k^u(t), \quad \delta_k(0) = 0, \\ \delta_k^y(t) = g_{kx}(w_k^0(t), u_k^0(t))\,\delta_k^x(t) + g_{ku}(w_k^0(t), u_k^0(t))\,\delta_k^0(t) \end{cases}$$

$$\tilde{I}_\epsilon: \begin{cases} w_k^0(t) = x_k^0(t) \\ u_k^0(t) = \gamma_k^0(y_0^0(t), u(t)) \\ \delta_k^u(t) = \gamma_{ky}^0(y_0^0(t), u(t))\,\delta_0^y(t) + \gamma_{k\epsilon}^0(y_0^0(t), u(t)) \\ \tilde{y}(t, \epsilon) = \sigma(y_0^0(t) + \epsilon\delta_0^y(t), u(t)) \end{cases}$$

where $y_0^0 \triangleq |y_1^{0\prime} y_2^{0\prime} \ldots y_N^{0\prime}|'$, $\quad \delta_0^y \triangleq |\delta_1^{y\prime} \delta_2^{y\prime} \ldots \delta_N^{y\prime}|'$.

Since $y_0(t, \epsilon) = y_0^0(t) + \epsilon\delta_0^y(t) + t(\epsilon)$ and σ is continuous, $\tilde{S}_\epsilon$ is apparently a first-order approximation of S_ϵ. The order of $\tilde{S}_\epsilon$ is twice the order of S_ϵ; however, note that:

(i) S_k' is linear for all $k = 1, 2, \ldots, N$;

(ii) contrary to S_ϵ, $\tilde{S}_\epsilon$ retains an extremely useful structural property of S_0 thanks to which the overall simulation can be carried on by simulating no more than a single subsystem at a time. In fact, the critical point, making such a decomposition of the simulation process possible, is the acyclicity of the system graph. Since the system graph G_0 associated with S_0 is, by assumption, acyclic a little thought reveals that the system graph $\tilde{G}_\epsilon$ associated with $\tilde{S}_\epsilon$ must be acyclic as well (while G_ϵ is not).

Singular Perturbation

The typical situation suited to be dealt with by singular perturbation techniques occurs when the dynamical system under consideration consists of several subsystems characterized by significantly different time-scales. To make the presentation as simple as possible, we shall assume that the system consists of two subsystems only: the fast and the slow one.

In engineering applications, the slow subsystem often accounts for the dynamics of those phenomena which are of primary importance to the application at hand, while the fast subsystem often describes 'parasitic' or secondary importance dynamic effects usually neglected in first approximation design. In this connection, it is important to notice that neglecting the fast dynamics does not at all mean neglecting the fast subsystem. Rather, it amounts to assuming that the fast subsystem is described by its steady-state equations. By no means, then, is neglecting the fast dynamics equivalent to assuming that the coupling between the slow and the fast subsystem is weak. For this reason, singular perturbation methods are also referrred to as '*strong coupling*' methods.

There are at least two good reasons for simplifying models by neglecting fast phenomena. One is that showing fast and slow transients at once makes the model 'stiff', in other words hard to handle (simulate) on a digital computer. A second one is that reducing the order of the model (or simplifying its interconnected structure) usually entails a much more than proportional reduction in the computational effort needed to analyse (or design) the system. However, especially in the design of high-quality control or decision systems, neglecting the fast dynamics may lead to quite unsatisfactory (even unstable) designs. The objective of the singular perturbation approach may then be twofold. First, it helps one to understand the circumstances under which the above simplification may become critical; second, it provides relatively cheap and effective ways of improving an oversimplified design, insisting on a 'piece-by-piece' rather than a 'single-shot' analysis (carried on looking at the original system as to a solid monolith). In what follows, we shall concentrate, for the sake of brevity, on the first aspect only, even though it is thanks to some results obtained along the second line of thinking that the singular perturbation approach has proved more fruitful and is therefore attracting the attention of an increasing number of researchers [8–12].

A singularly-perturbed two-timescale free dynamical system is typically described as follows:

$$S_\epsilon: \begin{cases} \dot{x}(t, \epsilon) = f(x(t, \epsilon), z(t, \epsilon), t, \epsilon), & x(0, \epsilon) = \xi(\epsilon), \\ \epsilon\dot{z}(t, \epsilon) = g(x(t, \epsilon), z(t, \epsilon), t, \epsilon), & z(0, \epsilon) = \zeta(\epsilon), \end{cases}$$

where x and z are the ν-dimensional state vector of the slow subsystem and the μ-dimensional state vector of the fast subsystem respectively; f, g, ξ and z are given functions; ϵ is a 'small' real parameter. Neglecting the fast dynamics of S_ϵ amounts to setting $\epsilon = 0$. We are therefore interested in studying the motion of S_ϵ, for ϵ in the neighbourhood of zero.

The *reduced system* S^0 is obtained from S_ϵ by setting ϵ equal to zero and dropping the initial condition for the fast state variable z, i.e.

$$S^0: \begin{cases} \dot{x}^0(t) = f(x^0(t), z^0(H, t, 0)), & x^0(0) = \xi(0); \\ 0 = g(x^0(t), z^0(H, t, 0)). \end{cases}$$

Assuming that a solution to S_0 exists and is unique, a first question is under what circumstances (and in what sense) the solution of S_ϵ exists, is unique and tends to the solution of S^0 as ϵ goes to zero, so that S^0 can be qualified in some proper sense as an approximation of S_ϵ. A second equally important question is under what additional conditions, possibly involving the fast subsystem only, does the stability of the motion of S^0 imply the stability of the motion of S_ϵ, at least for ϵ sufficiently small.

An answer to these questions can be obtained by analysing two additional systems involving the parameter ϵ in a regular way.

The *outer system* S_ϵ^* is, by definition, obtained from S_ϵ simply by substituting the initial condition of the fast variable z with the requirement that the motion of S_ϵ^* depends smoothly on ϵ in the neighbourhood of $\epsilon = 0$. More precisely:

$$S_\epsilon^*: \quad \begin{cases} \dot{x}^*(t, \epsilon) = f(x^*(t, \epsilon), z^*(t, \epsilon), t, \epsilon), & x^*(0, \epsilon) = \xi(\epsilon), \\ \epsilon \dot{z}^*(t, \epsilon) = g(x^*(t, \epsilon), z^*(t, \epsilon), t, \epsilon). \end{cases}$$

$x^*(t, \cdot)$, $z^*(t, \cdot)$ are continuously differentiable at $\epsilon = 0$, with $x^*(t, 0) = x^0(t)$ and $z^*(t, 0) = z^0(t)$, for all $t \geqslant 0$.

Now let $X \triangleq x - x^*$, $Z \triangleq z - z^*$ and $\tau \triangleq t/\epsilon$. With these new variables, S_ϵ becomes

$$\dot{X}(\tau, \epsilon) = \epsilon F(X(\tau, \epsilon), Z(\tau, \epsilon), \epsilon\tau, \epsilon), \quad X(0, \epsilon) = 0,$$
$$\dot{Z}(\tau, \epsilon) = G(X(\tau, \epsilon), Z(\tau, \epsilon), \epsilon\tau, \epsilon), \; Z(0, \epsilon) = \zeta(\epsilon) - z^*(0, \epsilon),$$

where $F(X, Z, t, \epsilon) \triangleq f(x^*(t, \epsilon) + X, z^*(t, \epsilon) + Z, t, \epsilon) - f(x^*(t, \epsilon), z^*(t, \epsilon), t, \epsilon)$ and G is defined in a similar manner. In this new form, S_ϵ seems to depend regularly on ϵ but considering the motion of S_ϵ on any however small finite interval $0 \leqslant t \leqslant \bar{t}$ calls, as ϵ goes to zero, for analysis over an indefinitely growing interval $0 \leqslant \tau \leqslant \bar{t}/\epsilon$. The new form of S_ϵ can therefore be effectively used to describe the behaviour of S_ϵ for t near zero, when ϵ goes to zero. Setting ϵ equal to zero yields $X(\tau, 0) = 0$ so that the fast subsystem can be rewritten as

$$S_{BL}(\theta): \quad \begin{aligned} &\dot{Z}(\tau; \theta) = g(x^0(\theta), z^0(\theta) + Z(\tau; \theta), \theta, 0), \\ &Z(0; \theta) = \zeta(0) - z^0(\theta), \end{aligned}$$

where θ has in fact to be treated as a nonnegative parameter.

$S_{BL}(\theta)$ is the *boundary layer* system associated with S_ϵ. It is intuitively appārent that in order for $(x(t, \epsilon), z(t, \epsilon))$ to converge to $(x^0(t), z^0(H))$, for every $t > 0$, as ϵ goes to zero, some stability properties of $S_{BL}(\theta)$ will be required. More specifically, note that convergence to $(x^0(t), z^0(t))$ for $t = 0$ is out of question, unless $z^0(0) = \zeta(0)$; furthermore, the origin $Z = 0$ is, by definition of x^0

and z^0, an equilibrium state of $S_{BL}(\theta)$; a typical requirement then will be that the origin be an asymptotically stable equilibrium state of $S_{BL}(\theta)$, for all $\theta \geqslant 0$, and that its domain of attraction at $\theta = 0$ is large enough to include $\zeta(0) - z^0(0)$. A precise result along this line, proved in [13] and [14], can be stated as follows.

THEOREM 5. Assume that:

(i) The functions f, g, ξ and z are bounded and twice continuously differentiable with respect to their arguments in some neighbourhood of $\{(x^0(t), z^0(t), t, 0), t \geqslant 0\}$, i.e. for $|x - x^0(t)|$, $|z - z^0(t)|$ and ϵ sufficiently small, for all $t \geqslant 0$.

(ii) All eigenvalues of $g_z(t) \triangleq \partial g/\partial z(x^0(t), z^0(t), t, 0)$ have real parts less than some fixed negative number, for all $t \geqslant 0$, so that the origin is an asymptotically stable equilibrium state of $S_{BL}(t)$.

(iii) The state $\zeta(0) - z^0(0)$ belongs to the domain of attraction of the origin in $S_{BL}(0)$.

(iv) The linear system $\dot{x}(t) = (f_x(t) - f_z(t)g_z^{-1}(t)g_x(t))x(t)$, where $f_x(t) \triangleq \partial f/\partial x(x^0(t), z^0(t), t, 0)$, etc. ... , is asymptotically stable so that (x^0, z^0) is an asymptotically stable motion of S^0.

Then, for each small $\epsilon > 0$, the following facts occur:

(A) S_ϵ admits a unique motion $(x(t, \epsilon), z(t, \epsilon))$; furthermore, there are an outer motion $(x^*(t, \epsilon), z^*(t, \epsilon))$ and a boundary layer motion $(X(t, \epsilon), Z(t, \epsilon))$ such that

$$x(t, \epsilon) = x^*(t, \epsilon) + X(t/\epsilon, \epsilon)$$
$$z(t, \epsilon) = z^*(t, \epsilon) + Z(t/\epsilon, \epsilon)$$

for all $t \geqslant 0$. In turn, the outer and the boundary layer solutions have a finite Mac Laurin expansion

$$x^*(t, \epsilon) = x^0(t) + \epsilon x^1(t) + 0(\epsilon^2), \quad X(\tau, \epsilon) = \epsilon X^1(\tau) + 0(\epsilon^2),$$
$$z^*(t, \epsilon) = z^0(t) + \epsilon z^1(t) + 0(\epsilon^2), \quad Z(\tau, \epsilon) = Z^0(\tau) + \epsilon Z^1(\tau) + 0(\epsilon^2),$$

where $0(\epsilon^2)$ holds uniformly for $t \geqslant 0$ as ϵ tends to zero; the functions x^1, z^1, X^1, Z^0 and Z^1 are determined by

$$S': \begin{cases} \dot{x}^1(t) = f_x(t)x^1(t) + f_z(t)z^1(t) + f_\epsilon(t), \quad x^1(0) = \dfrac{d\xi}{d\epsilon}(0), \\ z_1(t) = g_z^{-1}(t)(\dot{z}^0(t) - g_x(t)x^1(t) - g_\epsilon(t)); \end{cases}$$

$$\dot{Z}^0(\tau) = G(Z^0(\tau)), \qquad Z^0(0) = \zeta(0) - z^0(0),$$

$$\dot{X}^1(\tau) = F(Z^0(\tau)), \qquad X^1(0) = 0,$$

$$\dot{Z}^1(\tau) = G_X(Z^0(\tau))X^1(\tau) + G_Z(Z^0(\tau))Z^1(\tau) + G_\epsilon(Z^0(\tau)),$$

$$\begin{aligned} Z^1(0) &= \left[\frac{d}{d\epsilon}(\zeta(\epsilon) - z^*(0, \epsilon))\right]_{\epsilon=0} \\ &= -[g_z^{-2}(g_x f + g_t) + g_z^{-1}(g_x x^1 + g_\epsilon)]_{(x^0(0), z^0(0), 0, 0)} \end{aligned}$$

where, for any function $V \in \{F, G, G_T, G_z, G_\epsilon\}$, and any $Z \in R^\mu$, $V(Z)$ is a shorthand notation for $V(0, Z, 0, 0)$.

(B) The motion $(x(t, \epsilon), z(t, \epsilon))$ of S_ϵ is asymptotically stable and an estimate of the domain of attraction at $t = 0$ can be obtained by analysing the reduced system S^0 and the boundary layer system $S_{BL}(0)$, separately. □

As a final remark, we point out that, under the assumptions of Theorem 5, not only is S_0 indeed a zeroth order approximation of S_ϵ but, letting

$$\tilde{S}_{BL}: \quad \epsilon\dot{\sigma}^0(t, \epsilon) = G(\sigma^0(t, \epsilon)), \qquad \sigma^0(0, \epsilon) = \zeta(0) - z^0(0),$$

$$\tilde{S}'_{(BL)}: \quad \epsilon\dot{\chi}^1(t, \epsilon) = F(w^0(t, \epsilon)), \qquad \chi^1(0, \epsilon) = 0,$$

$$\tilde{S}'_{BL}: \quad \epsilon\dot{\sigma}_1(t, \epsilon) = G_X(w^0(t, \epsilon))w^1(t, \epsilon) + G_Z(w^0(t, \epsilon))\,\sigma^1(t, \epsilon) + + G_\epsilon(w^0(t, \epsilon)), \qquad \sigma^1(0, \epsilon) = Z^1(0),$$

$$\tilde{I}_\epsilon: \begin{cases} w^0(t, \epsilon) = \sigma^0(t, \epsilon) \\ w^1(t, \epsilon) = \chi^1(t, \epsilon) \\ \tilde{x}(t, \epsilon) = x^0(t) + \epsilon(x^1(t) + \chi^1(t, \epsilon)) \\ \tilde{z}(t, \epsilon) = z^0(t) + \sigma^0(t, \epsilon) + \epsilon(z^1(t) + \sigma^1(t, \epsilon)). \end{cases}$$

the interconnected composite system $\tilde{S}_\epsilon = (S^0, S', \tilde{S}_{BL}, \tilde{S}'_{(BL)}, \tilde{S}'_{BL}; \tilde{I}_\epsilon)$ is a natural candidate to be an improved approximation of S_ϵ. In this connection, note that each subsystem of $\tilde{S}_\epsilon$ is either of order ν, as the slow subsystem of S_ϵ, or of order μ, as the fast subsystem of S_ϵ, but, in addition, the system graph associated with $\tilde{S}_\epsilon$ is acyclic (so that $\tilde{S}_\epsilon$ can easily be solved sequentially subsystem-wise), whereas the system graph associated with S_ϵ is not. This feature alone, with all its possible consequences, may widely compensate for the drawback of $\tilde{S}_\epsilon$ being larger than S_ϵ, especially in view of the fact that all the subsystems of $\tilde{S}_\epsilon$ marked with a prime are in fact linear, by definition.

CURRENT TRENDS

Giving in a few lines a fairly balanced view of the current trends in such a wide and active research fields as interconnected systems analysis is an almost impossible task. Here are just a few, possibly biased, remarks.

A first refers to the somewhat intriguing relationship between design and analysis, as we have called it here in accordance with most of the existing literature on the subject. As is well known, especially in engineering practice, it is fairly customary to tackle hard and only partially formalized design problems by solving first a suitable set of (easier and) formally well specified analysis problems to be subsequently used in developing a rational trial-and-error searching procedure for an acceptable solution of the original (design) problem. This is a demanding job and a good deal of intelligence, intuition and knowledge is required to identify the problems of analysis for which a solution can be found transparent enough to make it useful also in design. Most of the current activity in each area of interconnected systems analysis is, in fact, design motivated and oriented. This is certainly true in Stability Theory, where for instance a relevant effort is spent in trying to devise new, rational, application-oriented methods for the design of robust, decentralized stabilising controllers. This is even more true in Singular Perturbation Theory where close consideration of challenging, long-standing and highly representative real design problems stimulates the development of a fairly sophisticated, design-oriented modelling machinery, the ultimate benefit of which is to give new and deeper insight into the almost completely open (design) problem of matching the structure of a (multiagent) decision-maker with those features only of the controlled system which are really relevant in the control problem at hand.

In the area of well-posedness and simulation of large-scale dynamical systems a major stimulus is provided by the rapidly expanding interest in Very Large Scale Integration (VLSI) technology for digital information processing. In particular, ever-growing is the need of faster, flexible dedicated simulation packages for the computer-aided analysis of VLSI circuits and, conversely, closer is coming the perspective of systematic, hierarchical computer-aided design procedures for ultra-fast VLSI simulators (digital realization) of large-scale dynamical systems, as they are needed in advanced filtering, prediction and control applications.

In decomposition-based stability theory, one of the most promising recently developed trends has been activated by the so-called Siljak's Inclusion Principle by which several ideas and design procedures based on standard decomposition (that is system decomposition induced by a partition of the state vector) are significantly improved by also incorporating overlapping decompositions, i.e. system decompositions in which any pair of subsystems may share one or more state variables.

The fundamental slow/fast decomposition idea of singular perturbation is currently spreading through a variety of seemingly weakly related fields such as modelling in networks and power systems, well-posedness of differential games, analysis of continuous-time stochastic systems, adaptive control, aggregation of Markov chains, implementability of algorithms, and advanced modelling in chemistry, biology and physics. Quite interesting is the recently demonstrated possibility of using singular perturbation techniques to deal with multiagent decision or control problems in which each agent has a different view, that is a 'personal' model of the rest of the world. We may conclude, then, by noting that, despite the conspicuous research work done in the past decades and some recent really relevant progress, there are still so many different, implicit, poorly understood situations modelled by a 'small' singular parameter affecting the behaviour of a dynamical system that even recognizing them and discovering the basic mathematical mechanism leading to multi-timescale dynamics remains an exciting challenge and a thoughtful activity. It is certainly a fundamental open field for further research.

ACKNOWLEDGEMENT

This work was supported by CNR-Centro di Teoria dei Sistemi, Italian MPI and Office of Naval Research, Contract no. 14–80–C–0507/ERL–UCB

REFERENCES

[1] M. Vidyasagar, 'On the Well-posedness of Large-scale Interconnected Systems', *IEEE Trans. on Automatic Control*, AC–25 (1980), 413–21.
[2] G. Guardabassi, 'Subsystemwise Simulation of Discrete-time Interconnected Dynamical Systems' (University of California, Dept. of Electrical Engineering and Computer Science, ERL M82/8, 1982).
[3] A. N. Michel and R. K. Miller, *Qualitative Analysis of Large-Scale Dynamical Systems* (Academic Press, New York, 1977).
[4] D. D. Šiljak, *Large-Scale Dynamic Systems: Stability and Structure* (North-Holland, New York, 1978).
[5] M. Araki, 'Stability of Large-scale Nonlinear Systems: Quadratic-order Theory of Composite-systems Method Using M-matrices', *IEEE Trans. on Automatic Control*, AC–23 (1978), 129–42.
[6] C. A. Desoer and M. Vidyasagar, *Feedback Systems: Input-Output Properties* (Academic Press, New York, 1975).
[7] E. J. Lasley and A. N. Michel, 'Input-output Stability of Interconnected Systems', *IEEE Trans. on Automatic Control*, AC–21 (1976), 84–9.
[8] R. E. O'Malley, Jr., *Introduction to Singular Perturbations* (Academic Press, New York, 1974).
[9] P. V. Kokotovic, R. E. O'Malley Jr., and P. Sannuti, 'Singular Perturbations and Order Reduction in Control Theory – An Overview', *Automatica* **12** (1976), 123–32.
[10] W. Eckhaus, 'Formal Approximations and Singular Perturbations', *SIAM Review* **19** (1977), 593.
[11] S. L. Campbell, *Singular Systems of Differential Equations* (Pitman, New York, 1980).
[12] V. R. Saksena and P. V. Kokotovic, 'Singular Perturbations in Control Theory: 1976–1982' (in press).
[13] F. Hoppensteadt, 'Properties of Solutions of Ordinary Differential Equations with Small Parameters', *Comm. on Pure and Applied Math.* **XXIV** (1971), 807–40.
[14] F. Hoppensteadt, 'Asymptotic Stability in Singular Perturbation Problems II: Problems Having Matched Asymptotic Expansion Solutions', *J. of Diff. Eq.* **15** (1974), 510–21.

Politecnico di Milano, Italy

G. Cohen

Optimization Techniques in Large-scale Systems

INTRODUCTION

From the point of view of real-life applications, optimization of large-scale systems is of great importance since improvement in the performance of such systems by even a few percent may involve savings of huge amounts of money. This is apparent when considering the cost of operating systems such as power or water-supply networks, for example. Unfortunately, these kinds of systems are very often characterized by the following features:

– *complexity*, that is non-homogeneity of characteristics such as timescales, degree of connectedness of each component with its neighbours, etc. . . . , and also nonlinear (and sometimes distributed parameter) equations, implicit dynamic equations of the form

$$f(x, \dot{x}, u) = 0$$

(x state variable, u control variable, $\dot{x} = dx/dt$, t time), usual in water supply networks, for example (see [9]);

– *large size* of the optimization problem in terms of number of equations, number of variables, number of constraints, and especially, in stochastic dynamic systems, number of state variables, etc. . . . ;

– *combinatorial* or *mixed* nature of the decision variables also very usual in several network applications (start-up policy of thermal generating units in power networks, on-off control of pumps in water supply networks, etc. . . .) or in investment optimization, due to the quantization of the size of the possible investment choices;

– *stochastic* nature of the operating conditions of such systems due to stochastic (power or water) demand, stochastic water inflows in hydropower systems, failures of components, etc. . . . .

Needless to say, in the face of such difficulties, we have no definite nor completely satisfactory answers. However, some of the above-mentioned features which are a source of difficulties may also be a source of simplification. For example, differences in timescales allow order-reduction or aggregation [15]. In networks, these differences in timescales may be caused by differences in degree (or intensity) of connectedness between components. For example, think of two water reservoirs strongly connected by a very wide, short pipe: if we take the sum of volumes of stored water in the reservoirs as the first state variable and the differences of water levels as the second one, then obviously the dynamics of the latter is very fast and can be neglected by setting this state variable to zero permanently, at least to get a first-order approximation of the optimal solution.

Viewing the whole system as being made up of interconnected subsystems, especially if the intensity or the number of interactions between subsystems is weak with respect to that of interconnections taking place in the subsystems, has been the motivation for a theory of *decomposition* in optimization (mathematical programming as well as deterministic optimal control). These interactions, although not relatively numerous nor relatively weak, cannot always be neglected, since they are sometimes an essential part of the management of the system. In water-supply networks, for example, they may represent exchanges of water between subnetworks, that represents a mutual help in facing locally difficult situations and a means of achieving an overall optimum. Hence, decomposition must be completed with *coordination*.

Decomposition-coordination algorithms will be discussed briefly in this chapter. Since they are generally iterative processes, they are mostly used for off-line computations, although some attempts for feeding them with on-line information have proved useful [5, 13]. Also the coordination variables (interactions, dual multipliers of coupling constraints) are generally found, in dynamic problems, as trajectories that is in open-loop form. As a matter of fact, trying to compute them in closed-loop form would pose the problem in which local or global information should be involved in the loop and, anyway, would be generally as difficult as solving the whole problem by dynamic programming, a task of course infeasible when the number of state variables exceeds three.

Nevertheless, an advantage of decomposition-coordination methods that has proved very significant in practical applications is that they enable the user to formulate independent or isolated subproblems. This opens the way to further simplications at the level of subproblem resolution, because aggregation or singular perturbation methods can more likely apply to the subsystems than to the whole system (think of our example of two strongly connected reservoirs), and, generally speaking, because one can more easily find ad-hoc simplications and well-suited or appropriate methods [1] when the problem is of medium or small size and presents some homogeneity. In this respect, decomposition and order-reduction or aggregation methods appear to be complementary ways of getting a solution for large-scale optimization problems. Moreover, in dynamic problems, it is only at the level of (possibly further simplified) subproblems that dynamic programming can be used, thus providing a local control in closed-loop form. Although not rigorously optimal out of the nominal trajectory to which the coordination iterative process has converged, it can be expected that this local feedback, by absorbing the local perturbations (at least partially) heuristically justifies keeping the coordination variables more or less invariant for small perturbations.

The possibility of using dynamic programming at the level of subproblem resolution may also help deal with the combinatorial aspect at this level, provided that the interactions are of a continuous nature so that duality and Lagrange multipliers still bear some significance. (see [9] for an instance in optimal control of water supply networks; a similar situation is encountered in the computation of the optimal start-up policy of thermal power generating units – the so-called unit commitment problem – where the only coupling constraint between a possibly large number of one-dimensional combinatorial dynamic problems is the satisfaction of the overall demand).

We have noticed that solving (dynamic) subproblems by dynamic programming provides local feedbacks around a trajectory which is overall optimal, thanks to coordination. These local feedbacks are not optimal but they are obtained through decomposed computations. There is another way to get local or *decentralized* feedbacks which are optimal in this class, but at the price of *global* computations which are generally even more costly than solving the problem in the class of *global* feedbacks. Hence one should clearly distinguish between *decomposition* of computations and *decentralization* of one-line informations. The latter requirement may be motivated by implementation constraints for systems widely distributed geographically, when the cost of centralization of information is too high compared with the loss of optimality caused by the restriction to local feedbacks only.

We shall discuss formulations that allow optimization in a class of feedbacks with structural constraints, but two remarks are already in order:

– one must be careful that the class of feedbacks contains at least all open-loop strategies, in order to make sure that the solution, which is obtained at the price of global computations, performs at least as well as solutions that could be obtained, say, as described before, by decomposition-coordination methods, in a deterministic setting of the problem;

– explicitly search for a solution in a class of feedbacks has no meaning in a purely deterministic setting, for there is an infinite number of representations of the same control history in terms of feedback strategies, and open-loop controls already allow one to achieve the best performance in this setting.

A TUTORIAL REVIEW OF DECOMPOSITION-COORDINATION METHODS IN OPTIMIZATION PROBLEMS

Although there are many formulations of optimization problems in which the basic methods can be demonstrated [2, 12, 16, 19], we will choose one which encompasses deterministic optimal control problems in a straightforward manner.

Consider the dynamics

$$(1) \qquad \dot{x}_i = \sum_{j=1}^{N} f_i^j(x_j, u_j), \qquad x_i(t_0) = x_{i0}; \qquad i = 1, \ldots, N$$

over a time horizon $[t_0, t_f]$, where $x_i(t) \in \mathbb{R}^{n_i}$ is the ith subsystem state and $u_i(t) \in \mathbb{R}^{m_i}$ its control at time t. Let us rewrite (1) in the following form

$$(2) \qquad \dot{x}_i = f_i^i(x_i, u_i) + v_i, \qquad x_i(t_0) = x_{i0},$$

where we have introduced the interaction input

$$(3) \qquad v_i \triangleq \sum_{j \in I_i} f_i^j(x_j, u_j),$$

where $I_i \triangleq \{1, \ldots, i-1, i+1, \ldots, N\}$.

Then (2) defines a causal relation of the form

$$(4) \qquad x_i = P_i(u_i, v_i),$$

where the symbols x_i, u_i, v_i stand now for trajectories (and not instantaneous values) and the initial condition x_{i0} is implicit. Using (4) in (3) allows to define a new relation

$$(5) \qquad v_i = \sum_{j \in I_i} h_i^j(u_j, v_j).$$

Similarly, considering the performance index

$$(6) \qquad G(u, x) = \int_{t_0}^{t_f} \left(\sum_{i=1}^{N} g_i(x_i, u_i) \right) \mathrm{d}t$$

one can define a new form, using (4) in (6),

$$(7) \qquad J(u, v) = \sum_{i=1}^{N} J_i(u_i, v_i),$$

where again the symbols u_i, v_i (and $u = (u_1, \ldots, u_N)$, $v = (v_1, \ldots, v_N)$), stand for functions of time. Finally, we consider the problem

$$(8) \qquad \operatorname*{Min}_{u,\, v} J(u, v), \text{ s.t.}^2 \ (5), u_i \in U_i, \quad i = 1, \ldots, N$$

where the feasible subsets U_i may represent any local constraints.

Price or Dual Decomposition

Considering multipliers p_i for the constraints (5), p_i lying in the dual space of v_i, we consider the ordinary *Lagrangian* functional of problem (8)

$$(9) \qquad L(u, v, p) = \sum_{i=1}^{N} [J_i(u_i, v_i) + \langle p_i, \sum_{j \in I_i} h_i^j(u_j, v_j) - v_i \rangle]$$

(where $\langle . , . \rangle$ denotes the duality product). Obviously we can re-write L in the form

$$(10) \qquad L(u, v, p) = \sum_{i=1}^{N} L_i(u_i, v_i, p).$$

Therefore, if L has a saddle-point, which is likely to occur under convexity assumptions, instead of solving problem (8), one can solve

$$(11) \quad \operatorname*{Max}_{p} \left[\operatorname*{Min}_{u,\, v} L(u, v, p), \text{ s.t. } u \in U_1 \times \ldots \times U_N \right],$$

where obviously the inner minimization problem splits up into N independent problems of the form

$$(12) \quad \operatorname*{Min}_{u_i,\, v_i} J_i(u_i, v_i) + \sum_{j \in I_i} \langle p_j, h_j^i(u_i, v_i) \rangle - \langle p_i, v_i \rangle, \text{ s.t. } u_i \in U_i,$$

where we have explicited the form of L_i. Of course, for any given 'price' vector p, the solution (u, v) produced by the resolution of the collection of problems (12) (that we assume unique) does not meet constraint (5). The coordination task amounts to updating p, e.g. by using the following rule (Uzawa algorithm)

$$(13) \quad p_i^{k+1} = p_i^k + \rho^k \left(\sum_{j \in I_i} h_i^j(u_j^k, v_j^k) - v_i^k \right),$$

where (u_i^k, v_i^k) corresponds to the solution of (12) with $p = p^k$, and ρ^k is a positive step, until (5) is approximately met.

Remark 1. Translating problem (12) back into the context of problem (1), (6), one can see that there are now additional (fictitious) decision variables v_i, a fact which might be, by intself, a drawback, in terms of computational load. Moreover, assume for example that f_i^i in (2) and g_i in (6) are linear in x_i for some i, then v_i will appear linearly in (12), which may lead to serious problems with a dual approach (see [10] for a discussion of this point). Such troubles are avoided with Augmented Lagrangians [3] [22] but unfortunately these Lagrangians do not share the separability property (10). However, decomposition methods using Augmented Lagrangians have been proposed [12, 13] and, more generally, the additivity property of cost (7) and constraints (5) is indeed not crucial for decomposition methods to work [6, 7].

Primal or Direct Coordination

This method is also called the *feasible* [4] or *resource allocation* [2] method. It consists in solving problem (8) in a reformulated version. Let us first introduce a new variable c made up of $N(N-1)$ elements $\{c_i^j; ((j \in I_i); i = 1, \ldots, N)\}$ where c_i^j lies in the same space as v_i. We shall also note

$$c_i = (c_i^1, \ldots, c_i^{i-1}, c_i^{i+1}, \ldots, c_i^N),$$
$$c^i = (c_1^i, \ldots, c_{i-1}^i, c_{i+1}^i, \ldots, c_N^i).$$

Then problem (8) is equivalent to

$$(14) \qquad \operatorname*{Min}_c \varphi(c) = \operatorname*{Min}_c \sum_{i=1}^{N} \varphi_i(c^i, c_i)$$

with $\varphi_i(c^i, c_i)$ being the optimal cost value of the following problem

$$(15\text{-}1) \quad \operatorname*{Min}_{u_i \in U_i} J_i(u_i, v_i)$$

s.t.

$$(15\text{-}2) \quad v_i = \sum_{j \in I_i} c_i^j$$

$$(15\text{-}3) \quad h_j^i(u_i, v_i) = c_j^i; \qquad j \in I_i.$$

Note that c^i appears as the right-hand side of the collection of constraints (15-3) while c_i appears in the definition (15-2) of v_i. Note also that once the N subproblems (15) have been solved for any given c, constraints (5) are met, hence the name 'feasible' for the method.

The coordination task amounts to solving problem (14). The functional φ is generally nondifferentiable and various methods have been proposed to solve (14), based on approximations of φ by upper hulls of an increasing number of linear functionals [16], or on subgradient iterations to update c [20]. Note that a subgradient of φ is given by the following formula

$$(16) \qquad \partial_{c_i^j}\varphi(c) = \nabla_{v_i} J_i(u_i, v_i) + \sum_{k \in I_i} \left[\frac{\partial h_k^i}{\partial v_i}(u_i, v_i) \right]^* p_k^i - p_j^i,$$

where the left-hand side denotes a partial subgradient with respect to c_i^j, $\nabla_{v_i} J_i$ denotes the partial gradient of J considered as a column vector (or primal element), p_j^i is an optimal multiplier of constraint (15-3), [ ]* denotes the adjoint (transpose in finite dimensional spaces) of a linear operator, and all arguments (u_i, v_i) are solutions of (15) for the given c.

Remark 2. Suppose $\dim u_i = m_i$, $\dim v_i = n_i$. In (15-3), there are $\Sigma_{j \in I_i} n_j$ scalar constraints for m_i free decision variables in (15-1). In many problems, it will happen that there are more constraints than decision varibles in (15) which, hence, will be unfeasible. This is the price for the method to be 'overall feasible', as discussed earlier. The situation is generally hopeless in dynamic problems since translated back into notations of (3), (15-3) would be

$$f_j^i(x_i(t), u_i(t)) = c_j^i(t), \qquad \forall t \in [t_0, t_1].$$

Even if we assume $m_i > \Sigma_{j \in I_i} n_j$ for finite dimensional problem with scarce interactions, it may happen that, for some values of c, the feasible set of (15) be empty due to the interplay of (15-3) with the constraints $u_i \in U_i$ (or even of (15-2) and (15-3) with constraints $(u_i, v_i) \in W_i$ that may arise if there are constraints initially imposed on x_i – see (4)). Selecting only values of c which do not cause this emptiness is generally not an easy task. For these reasons, it has been proposed to replace constraints such as (15-3) by additional *penalty* terms in the cost function [13].

Interaction Prediction Method

This method was originally proposed by Mesarovic *et al.* [19]. With respect to the primal method, constraints (15-3) are relaxed, thus avoiding the troubles described in Remark 2. Consequently, constraints (5) are met only once the coordination process has converged, as it is the case with the price method. For the same reason, additional terms such as those appearing in (12) must be incorporated in the local performance indices. Note however that, unlike (12), the local minimizations will now be with respect to the 'natural' decision variables u_i only, while v_i will be fixed (or predicted, hence the name of the method) by the coordination level, thus avoiding the troubles mentioned in Remark 1.

Namely, at step k of the coordination process, subproblem i will be

$$(17) \qquad \operatorname*{Min}_{u_i \in U_i} J_i(u_i, v_i^k) + \sum_{j \in I_i} \langle p_j^k, h_j^i(u_i, v_i^k) \rangle$$

whose solution is noted u_i^k, while the new coordination variables v^{k+1} and p^{k+1} for the next iteration will be computed by

$$(18) \qquad v_i^{k+1} = \sum_{j \, \mathrm{B} \, I_i} h_i^j(u_j^k, v_j^k); \qquad p_i^{k+1} = \nabla_{v_i} J_i(u_i^k, v_i^k).$$

Formula (18) can be obtained several ways (e.g. see [11]), but one can check that overall optimality conditions are met by the subproblem solutions, if convergence is assumed (i.e. $u_i^{k-1} = u_i^k, \ldots$). Convergence conditions have been studied by Cohen [7] for linear constraints and quadratic performance index.

Remark 3. One may think of replacing v^k by v^{k+1} in the right-hand side of Equation (18). This requires solving an implicit equation and in the context of optimal control problems this amounts to integrating the *global* differential Equation (1) together with its adjoint state (backward) equation, for the computation of p^{k+1}. This more complicated algorithm has, however, nice properties of adaptation in case of subproblem controller failures, when used on-line (see [5] for details).

Remark 4. If coupling constraints do not have the separable form (5), that is $v_i = h_i(u, v)$, then the second term in (17) can be replaced by

$$\sum_{j=1}^{N} \langle p_j^k, \nabla_{u_i} h_j(u^{k-1}, v^k) \,.\, u_i \rangle .$$

Similar considerations also apply to the cost function, allowing one to relax all additivity assumptions [6].

Remark 5. As already discussed, almost all the troubles mentioned in Remarks 1 and 2 disappear with this method. One may persist in the case where there are constraints $(u_i, v_i) \in W_i$, caused by constraints on x_i (given by (4)), for the set of u_i's meeting $(u_i, v_i^k) \in W_i$ might be empty for some v_i^k. A possible remedy, in this case, can be to use *penalty* or *exact penalty* techniques for taking care of these local constraints.

A General Theory of Decomposition-Coordination Algorithms

Cohen [6, 7] proposed the so-called Auxiliary Problem Principle as a general framework for decomposition-coordination algorithms. Consider the abstract optimization problem

$$\text{(19)} \qquad \operatorname*{Min}_{u} J(u) \text{ s.t. } u \in U_1 \times \ldots \times U_N,$$

where it is meant that the feasible set is decomposed (no interaction through constraints). Instead of solving (19) one can solve the sequence of auxiliary problems (u^k being currently known)

$$(20)\ \underset{u}{\mathrm{Min}}\, K(u) + \langle \epsilon^k J'(u^k) - K'(u^k), u \rangle \text{ s.t. } u \in U_1 \times \ldots \times U_N,$$

where K is an auxiliary functional, ϵ^k is a positive number (J' and K' denote derivatives). Let u^{k+1} be the solution of (20). If we assume that this iterative process is pursued until convergence is reached ($u^k \approx u^{k+1}$ - see [7] for convergence analysis), it is easy to prove that the limit of u is a solution of (19). Now since K is to be chosen by the user, one can choose an additive K, namely

$$K(u) = \sum_{i=1}^{N} K_i(u_i)$$

so that (20) splits up into N independent subproblems. Note that the key idea of decomposition appears here to be the linearization of J (which initially introduced the coupling in (19), for we do not assume it additive) since J appears in (20) only through its derivative. Extending the auxiliary problem principle to saddle-point problems, one is able to deal with coupling arising from constraints too, by using duality. Most of the existing decomposition algorithms can be embedded in this framework [6]. However, the main advantages of this approach are to show that separability is not an absolute requirement for decomposition, it also allows one to embed classical and decomposition algorithms in the same framework (e.g. choosing $K(u) = \frac{1}{2}\|u\|^2$ allowing one to recover a projected gradient algorithm through (20)), and finally it provides systematic guidelines for designing new algorithms and analysing their convergence.

Although differentiability (that is linearity of the directional derivative with respect to the direction) initially appeared as a key instrument for decomposition, the theory of [6, 7] has now been extended to nondifferentiable optimization problems too [25]. This opens up new possibilities such as using exact penalty techniques [3] to deal with coupling constraints; this may be helpful for nonconvex problems when duality gaps are encountered with price methods.

A BRIEF ACCOUNT OF DECENTRALIZED OPTIMAL CONTROL

For reasons that have been advocated in the introduction, it may be necessary, for some systems, to implement a closed-loop control strategy with decentralization of information, that is each control input (or group of control inputs) may only use a particular subset of the measurements. This problem, however, has no meaning in a purely deterministic situation since open-loop strategies are already optimal. In a stochastic framework, and in the context of optimization, the use of so-called 'nonclassical information patterns' is a source of enormous difficulties, as shown by Witsenhausen's counterexample [24]. For this and other reasons discussed by Benveniste *et al.* [1], it is therefore reasonable not only to define the information pattern (i.e. which measurement is to be used by which control input) but also to restrict the controller structure, e.g. by letting it belong to a parametrized class of controllers meeting the specified information pattern.

We shall discuss this approach in the so-called LQG framework (linear dynamics, quadratic cost function, Gaussian white noises) but Quadrat [21] has shown how this also extends to the nonlinear stochastic case.

Since the LQ optimization approach is also a way of synthesizing stabilizing feedbacks, we shall first briefly discuss this problem of stabilizing a system by decentralized control. Of course, the first question that arises in this case is whether or not such a problem has a solution, or, equivalently, which poles of the system can be moved by decentralized static output feedback or decentralized dynamic compensation. The reader is referred to Wang and Davison's [23] and other related papers for algebraic system theoretical answers.

More in the spirit of quadratic synthesis, one may refer to Medanic's paper [18] which is based on properties of an extended algebraic Riccati equation. This and many other similar works consider the deterministic linear system, and the stochastics, when explicitly introduced, are put on the initial condition, for the problem to make sense. The solution of the formulated LQ optimization problem generally depends on this initial probability law. Medanic [18] has analysed how this arbitrariness affects the poles of the optimal complete state feedback controlled system when passing to the decentralized output feedback system.

We now consider the LQG optimization problem while assuming that the cost function has a real meaning (that is optimization is not an indirect way for stabilizing the system; to mark this, we consider the finite time horizon problem).

Remark 6. In the stationary infinite time horizon problem, the possibility of stabilizing via decentralized control is a necessary condition for well-posedness of the problem. This is discussed in the above references. □

Consider the linear stochastic differential equations

(20) $\mathrm{d}x = (Fx + Gu)\,\mathrm{d}t + \mathrm{d}v,$

(21) $\mathrm{d}y = Hx\,\mathrm{d}t + \mathrm{d}w,$

where (20) is the dynamics relating the control input $u(t) \in \mathbb{R}^m$ and the dynamics 'noise' $v(t) \in \mathbb{R}^n$ to the state $x(t) \in \mathbb{R}^n$, F, G being (possibly time-dependent) matrices of appropriate dimensions, (21) is the observation equation relating the state and the observation 'noise' $w(t) \in \mathbb{R}^p$ to the observation $y(t) \in \mathbb{R}^p$; v and w are independent Wiener processes with

$$\mathbb{E}\{v(t)\} = 0, \quad \mathbb{E}\{v(t)v'(s)\} = V|t-s|,^4$$
$$\mathbb{E}\{w(t)\} = 0, \quad \mathbb{E}\{w(t)w'(s)\} = W|t-s|$$

[14]. The initial condition on $x(t_0)$ is a random Gaussian vector. We assume

$$(22) \quad \mathbb{E}\{x(t_0)\} = 0, \quad \mathbb{E}\{x(t_0)x'(t_0)\} = \Sigma_0.$$

Also $y(t_0) = 0$. The cost function one would like to minimize is

$$(23) \quad \mathbb{E}\left\{\frac{1}{2}\int_{t_0}^{t_f} (x'Qx + u'Ru)\,\mathrm{d}t\right\},$$

where Q (resp. R) is a symmetric nonnegative (resp. positive) matrix of appropriate dimension. The class of admissible closed-loop strategies are defined as feedbacks

$$(24) \quad u = C\zeta$$

on the output of a 'filter'

$$(25) \quad \mathrm{d}\zeta = A\zeta\,\mathrm{d}t + B\,\mathrm{d}y, \quad \zeta(t_0) = 0,$$

where $\zeta(t) \in \mathbb{R}^n$. Note immediately that if $r = p$, $A = 0$, $B = I$, then z can be identified with y and (24) is an output feedback. In general, the strategy (24)–(25) embeds all linear strategies including combinations of output feedbacks and dynamic compensation. With $r = n$ and all entries of A, B, C free for tuning, we are in the framework of classical information pattern.

As an example of decentralized information pattern, assume the term Gu in (20) is a sum of terms $\Sigma_{i=1}^{N} G_i u_i$, where $u_i(t) \in \mathbb{R}^{m_i}$

is the control input of controller i and (21) is made up of N observation equations relating x and w (or possibly w_i) to the 'private' observations y_i. Similarly, N independent 'filters' (A_i, B_i, output z_i, input y_i) are built as in (25) and N feedbacks $u_i = C_i\zeta_i$ are used (see (26)). This can be translated, in the general setting, by saying that the composite matrices A, B, C, which include all the design parameters, are constrained to be block-diagonal (with respect to block-decompositions of u, y, z into subvectors). Hence one can see that many information patterns can be specified simply by specifying which entries of A, B, C are available to the designer (possibly under constraints put upon their values) and which are fixed by the pattern to null (or nonnull) values.

To further simplify and condense the notations, one can consider the composite equation, instead of (20)–(21)–(24)–(25)

$$(26)\qquad \mathrm{d}\begin{pmatrix} x \\ \zeta \end{pmatrix} = \begin{pmatrix} F & GC \\ BH & A \end{pmatrix}\begin{pmatrix} x \\ \zeta \end{pmatrix}\mathrm{d}t + \begin{pmatrix} I & 0 \\ 0 & B \end{pmatrix}\mathrm{d}\begin{pmatrix} v \\ w \end{pmatrix},$$

where some of the entries in the composite matrices are chosen (as functions of time) so as to minimize the cost function

$$(27)\qquad \mathbb{E}\left\{\frac{1}{2}\int_{t_0}^{t_f} (x'Qx + \zeta'C'RC\zeta)\,\mathrm{d}t\right\},$$

(reformulation of (23)–(24)), where the expectation is taken according to the initial distribution of $(x', \zeta')'$ and the Wiener process laws.

The solution of such problems has been considered in several works (e.g. see [17]). Essentially, the cost function (27) can be expressed as a deterministic integral function of the time-dependent covariance matrix of the vector $(x'(t), \zeta'(t))'$. This covariance matrix obeys a deterministic differential equation. Hence, one comes up with a deterministic optimal control problem in matrix form, where the control inputs are the available entries in A, B, C above. Using Pontryagin's minimum principle one can write down necessary optimality conditions under the form of a two-point boundary value problem plus stationarity conditions of the Hamiltonian. Of course this requires introducing costate variables under the form of a symmetric matrix whose interpretation will be discussed below.

To avoid complex equations, let us consider the simplified situation where one wants to control (20) with decentralized state feedback $u_i = C_i x_i$ (assuming $x' = (x_1', \ldots, x_N')$) while optimizing (23). The reader will see how to generalize the following to the composite case (26)–(27) in a straightforward manner.

Let C denote the block-diagonal feedback matrix made up of the C_i's; one has, using the properties of the trace (noted 'tr'), that (23) is now equal to

$$\text{(28)} \qquad \operatorname{tr} \mathbb{E} \left\{ \frac{1}{2} \int_{t_0}^{t_f} (Q\Sigma + C'RC\Sigma) \, \mathrm{d}t \right\},$$

where $\Sigma(t) = \mathbb{E}\{x(t)x'(t)\}$ obeys the Lyapounov equation

$$\text{(29)} \qquad \dot{\Sigma} = (F + GC)\Sigma + \Sigma(F' + C'G') + V, \qquad \Sigma(t_0) = \Sigma_0.$$

The problem is thus to minimize (28) with respect to block-diagonal C for Σ driven by (29). In (29), there are only $n(n+1)/2$ independent scalar differential equations since Σ is symmetric. However, one can as well consider nonsymmetric initial conditions Σ_0 from the mathematical viewpoint, even if there is no practical interpretation. Then there are n^2 state variables and one must introduce n^2 costate variables arranged in an $n \times n$ matric Λ. Afterwards, once optimality conditions have been derived, one can realize that, with symmetric Σ, only $P = \Lambda + \Lambda'$ is involved in the equations. The detailed calculations are shown in [8]. Here we shorten the presentation, which is therefore mathematically incorrect, by introducing a symmetric costate matrix P directly, while using differentiation rules with respect to nonsymmetric matrices.

The Hamiltonian is

$$\mathscr{H} = \operatorname{tr}\{(Q + C'RC)\Sigma + P[(F + GC)\Sigma + \Sigma(F' + C'G') + V]\}. \tag{30}$$

Using the rules of matrix differentiation [8] and Pontryagin's minimum principle, one gets the necessary optimality conditions consisting of (29) and

$$\dot{P} = -P(F + GC) - (F' + C'G')P - Q - C'RC, \qquad P(t_f) = 0, \tag{31}$$

$$[(RC + G'P)\Sigma]^d = 0, \tag{32}$$

where $[\;]^d$ means 'block-diagonal part' with respect to given decompositions (here, of u and x).

Remark 7. As already explained in the introduction, it is reasonable to make sure that open-loop controls are included in the admissible set of strategies. In our case, this would be obtained by considering strategies of the form $u = Cx + h$, where h is a function of time. However, if $\mathbb{E}x(t_0) = 0$ and if there are no linear terms in the final and integral cost function (23) nor external deterministic input into the dynamics (20), then one can prove that $h = 0$. Otherwise, the optimality conditions should be extended [8] and one can see that the optimal open-loop part, or in other words the optimal distribution of the control effort between open-loop and feedback, is not independent of the noise and initial condition statistics V, W, Σ_0. This is a general phenomenon even when the open-loop part is null. It is revealed by (32) where C appears to be a function of Σ. There are exceptions to this rule, the most commonly known being the full state feedback situation. As a matter of fact, if any C is allowed in (32) (which thus must be considered without the symbol $[\;]^d$), then by taking $C = -R^{-1}G'P$, C is now only a function of P which does not depend on Σ (see (31)). This is the *certainty equivalence principle* (the optimal feedback matrix of the stochastic problem is the same as that of the deterministic one). There are some other special situations where such a phenomenon arises but they are exceedingly rare. □

There are two ways of looking at the same problem, which of course lead to the same optimality conditions but which provide an interpretation of matrix P. Namely, let us use the control law $u = Cx$ in (20) and let us compute the cost function (27), starting the integration at some $t_i \in [t_0, t_f]$ and assuming that the initial distribution of x at this time is a Gaussian distribution with null first moment and second moment equal to Σ. Then the expectation of this cost is given by the following expression

$$(33) \qquad \operatorname{tr} \frac{1}{2} \{P(t_i)\Sigma\},$$

where $P(t)$ is given by the backward integration of Equation (31). Hence the problem of minimizing (23) with respect to feedback matrices C under structural constraints can also be viewed as that of minimizing (33) for $t_i = t_0$, $\Sigma = \Sigma_0$, where P obeys the (backward) state Equation (31). Then (29) turns out to be the (forward) costate equation corresponding to the (initial) cost function (33).

From the numerical viewpoint, several methods of resolution can be considered, ranging from gradient [8] or congugate gradient methods (the gradient of the cost with respect to C is given by the left-hand side of (32)), to decomposition-coordination or block-relaxation methods [17]. In the latter case, although a block-diagonal C contains N nonnull blocks, the other matrices involved (P, Σ) contain $N(N+1)/2$ blocks, so that there are potentially $N(N+1)/2$ subproblems. However, one cannot give an interpretation of these subproblems in terms of suboptimizations as it was the case with the decomposition methods discussed earlier.

NOTES

[1] Note that each subproblem can be solved by a different method.
[2] 'subject to'.
[3] They also help to overcome duality gaps (non existence of saddle-points).
[4] The accent denotes 'transposition', $\mathbb{E}$ denotes 'mathematical expectation'.

REFERENCES

[1] A. Benveniste, P. Bernhard, and G. Cohen, 'On the Decomposition of Stochastic Control Problems', in G. Guardabassi and L. Locatelli (eds.), *Proceedings of 1st Ifac Symposium on Large Scale Systems Theory and Applications*, Udine, Italy (Pergamon Press, London, 1976).

[2] P. Bernhard, *Commande optimale, décentralisation et jeux dynamiques* (Dunod, Paris, 1976).

[3] D. P. Bertsekas, 'Necessary and Sufficient Conditions for a Penalty Method to Be Exact', *Mathematical Programming* **9** (1975), 87–99.

[4] C. B. Brosilow, L. Lasdon, and J. D. Pearson, 'Feasible Optimization Methods for Interconnected Systems', *Proceedings of Joint Automatic Control Conference* (Troy, N.Y., 1965).

[5] G. Cohen, 'On an Algorithm of Decentralized Optimal Control', *Journal of Mathematical Analysis and Applications* **59** (2) (1977).

[6] G. Cohen, 'Optimization by Decomposition and Coordination: A Unified Approach', *IEEE Transactions on Automatic Control*, AC–23 (2) (1978).

[7] G. Cohen, 'Auxiliary Problem Principle and Decomposition of Optimization Problems, *Journal of Optimization Theory and Applications* **32** (3) (1980).

[8] G. Cohen, 'Commande par contre-réaction sous contrainte de structure du gain d'un système stochastique linéaire. Résolution par une méthode de gradient', French journal *RAIRO*, Yellow Series, J-3 (1975), 33–46.

[9] G. Cohen, 'Optimal Control of Water Supply Networks', in S. G. Tzafestas (ed.), *Optimization and Control of Dynamic Operational Research Models* (North-Holland, Amsterdam, 1982).

[10] G. Cohen and J. F. Balducchi, 'A New Algorithm in Nondifferentiable Optimization and Its Use for Computing Saddle-points', *Proceedings of 18th Allerton Conference on Communication Control and Computing* (1980).

[11] G. Cohen and G. Joalland, 'Coordination Methods by the Prediction Principle in Large Dynamic Constrained Optimization Problems', in volume cited in ref. [1].
[12] G. Cohen and D. L. Zhu, 'Decomposition Methods in Optimization Using Augmented Lagrangians', presented at *Optimization Days*, Montréal, Canada, 1982 (in press).
[13] W. Findeisen, F. N. Bailey, M. Brdyś, K. Malinowski, P. Tatjewski, and A. Wozniak, *Control and Coordination in Hierarchical Systems* (Wiley, London, 1980).
[14] N. H. Fleming and R. W. Rishel, *Deterministic and Stochastic Optimal Control* (Springer-Verlag, Berlin, 1975).
[15] P. V. Kokotovic, R. E. O'Malley Jr., and P. Sannuti, 'Singular Perturbations and Order Reduction in Control Theory: An Overview', *Automatica* **12** (1976), 321–5.
[16] L. S. Lasdon, *Optimization Theory for Large Systems* (Macmillan, London, 1970).
[17] D. P. Looze and N. R. Sandell Jr., 'Decomposition of Linear, Decentralized Stochastic Control Problems', in A. Titli and M. G. Singh (eds.), *Proceedings of Ifac Workshop on Control and Management of Integrated Industrial Complexes*, Toulouse, France (Pergamon Press, London, 1977).
[18] J. Medanic, 'Synthesis of Decentralized Output Regulators', *Proceedings of Systems Engineering for Power: Organizational Forms for Large Scale Systems*, Davos, Switzerland, Vol. II (U.S. Dept. of Energy, Washington D.C., 1979).
[19] M. D. Mesarovic, D. Macko, and Y. Takahara, *Theory of Hierarchical, Multilevel Systems* (Academic Press, New York, 1970).
[20] B. T. Polyak, 'Subgradient Methods: A Survey of Soviet Research', in C. Lemaréchal and R. Mifflin (eds.), *Non-smooth Optimization*' (Pergamon Press, London, 1978).
[21] J. P. Quadrat, 'On Optimal Stochastic Control for Large Systems', in N. H. Fleming and Gorotsitza (eds.), *Recent Advances in Stochastic Control and Nonlinear Filtering*, Conference held at Cocoyoc, Mexico (Springer-Verlag, New York, 1982).
[22] R. T. Rockafellar, 'A Dual Approach to Solving Nonlinear Programming Problems by Unconstrained Optimization', *Mathematical Programming* **5** (1973), 354–73.
[23] S. H. Wang and E. J. Davison, 'On the Stabilization of Decentralized Control Systems', *IEEE Transactions on Automatic Control*, Vol. AC–18 (5) (1973).
[24] H. S. Witsenhausen, 'A Counter-example in Stochastic Optimal Control', *SIAM Journal of Control* **6** (1) (1968).

[25] D. L. Zhu, *Optimisation non différentiable et méthodes de décomposition*, Thesis dissertation, Ecole Nationale Supérieure des Mines de Paris, Centre d'Automatique et Informatique (Fontainebleau, France, 1982).

CAI-ENSMP, Fontainebleau and INRIA, Le Chesnay, France

S. Rinaldi

Systems Analysis and Environmental Modelling

INTRODUCTION

The aim of this article is the presentation of research activities in the field of Environmental Systems Analysis. In particular, emphasis will be given to the research areas which are more likely to be developed during the next few years. I hope that this contribution will be of help to managers of research institutions, and consulting groups as well as to graduate students, Ph.D. Candidates and, in general, to people who will be concerned with this discipline in the future.

Environmental systems analysis is very much diversified, both as regards sectors (natural resources, pollution, geology, land use, ecology, etc.) and problems (simulation, forecast, management, and so on). Thus, it is quite difficult to offer a synthetic presentation of all the material. Moreover, the level of research and the effort planned for this field of science in the future vary significantly from one country to another, so that it is important to stress that the general statements made in what follows refer to the most developed countries. Consistently, the conclusions and guidelines given at the end of the article must be considered with remarkable caution before using them in planning the activities of consulting firms or research institutes. This is particularly true if considerations of short-term 'productivity' have to be taken into account.

The whole area may be subdivided into the following five sectors:

– water resources;
– land use;
– air pollution;
– water pollution;
– ecology and renewable resources.

This classification is certainly not exhaustive, since important sectors (e.g. geology and meteorology) are missing. Nevertheless, it fits with my own scientific background [1] and the experience gained while organizing and coordinating a national research program [2] and two international congresses [3, 4]. The way the problems are classified, namely

– simulation;
– planning;
– forecasting;
– management;

is also rooted in my past experience but is, nevertheless, widely accepted by the scientific community.

Some of the statements made in the following are probably a consequence of two long visits to the University of Kyoto (Japan) and at the University of British Columbia (Vancouver, Canada) where I worked respectively on air pollution modelling in 1974–75 and renewable resources management in 1976. Moreover, my experience on river pollution modelling and water resources management at the International Institute for Applied Systems Analaysis (Laxenburg, Austria) in 1977, 1978 and 1981, has also a great deal in common with the content and conclusions of this article.

In the following sections the five sectors mentioned above will be analysed in the order. In doing so, the problems of simulation, planning, forecasting, and management will be kept separate, so that the relative importance of the different methodologies in all sectors can clearly emerge. In each section, I will give some useful reference to the most important sources of information (journals, congresses and schools) and to the most outstanding research institutions. In the final main section a comparison of all sectors and problems will allow us to draw some synthetic conclusions about the most likely future development of teaching and research activities in this fascinating field.

WATER RESOURCES

The majority of the studies carried out in the field of water resources are devoted to the temporal and spatial allocation of the resource to a set of 'users'. Within this general definition we can find a great variety of cases and subcases, depending upon the economic, temporal, and spatial scale of the problem. The first significant work in the field was probably the Harvard Water Resources Program [5], which was developed in the early sixties and has strongly influenced most subsequent river basin planning and management studies. Since then, relevant progress has been achieved from a theoretical and conceptual point of view by extensively and properly applying optimization techniques and control theory. On the other hand, important realizations have been performed with the aid of modern technologies for data collection and analysis. For example, in the Columbia River Basin (almost 650,000 km^2) a network of computers is used permanently to store all the hydrometeorological data, to develop short-, medium-, and long-term forecasts and to supply many local agencies in the system with the most appropriate suggestions for their decision-making.

Simulation

Numerous simulation models were proposed and tested, particularly in the period 1960–70. Such models dealt with the various components of the water cycle (relationships between precipitation, surface runoff, and infiltration, flood propagation, snow melt, lake circulation, evaporation, etc.). Many of these models, which have their origins in very old work such as the equation of Barré de Saint Venant published in 1871, are available today in the form of computer codes supplied together with suitable user manuals [6]. Research in this field is declining and limited to very technical questions like convergence of parameter estimation algorithms and numerical schemes for the solution of partial differential equations. Nevertheless, a few specific topics like, for example, 'solid transport' [7] and 'yield response to water' [8] certainly deserve some attention.

Planning

Linear and non-linear programming models have been used extensively to solve significant problems such as the determination of the number and capacities of artificial reservoirs to be built in a river basin, and the design of a hydraulic network needed to distribute the resource among different agricultural, industrial, and civil users [9, 10]. In such studies, major considerations are the economic aspects of the problem [11, 12].

Forecasting

The most interesting research topic is the real-time forecasting of river flow rates, which is of enormous economic and social impact, especially where floods are concerned. Time series analysis [13] and Kalman filtering [14] have been used extensively and have proved to be valuable tools for supplying reliable forecasts. Other topics, such as the prediction of water demand in the residential [15], agricultural, and industrial [16] sectors have also been explored to a certain extent.

Management

The problem of real-time operation of one or few reservoirs has been thoroughly analysed in the past (see, for instance [17]). All possible techniques (linear programming, deterministic and stochastic dynamic programming, non-linear programming, maximum principle and game theory) have been successfully applied to real or realistic situations, so that satisfactory and general results are now available for the management of single reservoir systems. Nevertheless, the problem is still open to speculation in the more complex case of multireservoir systems. In particular, the real-time coordination of interconnected multipurpose reservoirs represents the key point of a number of research programmes currently under investigation.

There is no doubt that the use of systems analysis in the field of water resources has been very productive. On a conceptual (and hence didactic) level the main contribution has been the possibility of developing a simple and unified approach to the various dynamic phenomena characterizing the water cycle. Moreover, for some simple but significant cases, like the so-called linear reservoir problem, it has been possible to develop a complete and comprehensive theory mainly based on linear systems theory and optimal control. On the other hand, large-scale problems have also been solved via simulation, and various techniques for automatic data analysis have been designed and implemented by a number of river basin authorities.

The progress of theoretically oriented research during the last twenty years has been such that one can expect for the future a great effort in applications (case studies). Nevertheless, management and coordination problems will still deserve some attention from a conceptual point of view. Moreover, almost unexplored are the problems of conflict analysis and resolution in multidecision-maker systems, and of reliability of the optimal solutions. A related topic which still needs a relevant effort concerns the visualization of the results and, in general, the direct interaction between the computer models and the users.

Research on these topics is presently carried out by many research institutes, among which we might list the US Hydrologic Engineering Center (Davis, California, USA), the Department of Civil Engineering of the University of Colorado (Fort Collins, Colorado, USA), the Institute of Hydrology (Wallingford, UK), and the Center for Large Scale Systems and Policy Analysis of the Case Western Reserve University (Cleveland, Ohio, USA), which organizes each year an international course on water resources management and planning.

The most important contributions in this field can be found in the journals *Water Resources Research, Advances in Water Resources, Journal of Hydrology*, and *Applied Mathematical Modelling* and in the publications of the International Association of Hydrological Sciences.

LAND USE

The problem of description and control of social equilibria has been considered of great relevance ever since the last century, and during this time formal models have been employed to analyse phenomena which are characterized by significant geographical variations. After the first investigations on the agricultural sector carried out by Von Thünen in 1826, we might recall the studies of Launhardt (1885) on the siting of industrial activities, and the first optimization models due to Weber (1909). Since then, a huge amount of work has been done on city and regional planning, first in Germany (Lösch, Christaller) and then (in the early fifties) in the United States (Isard [18] and Lowry) and in the United Kingdom. Indeed, the most significant results have been obtained in the United States with a series of models concerning the urban system (traffic, urban development and services) and in the UK with numerous experiences of urban and regional planning during the early seventies.

Simulation

The development and the dynamics of urban and regional systems have been analysed by means of various descriptive models. Among them, we can recall the models for land use [19], and those for spatial location, for example, Lowry's model [20], and traffic (in particular the well-known models of trip generation, trip distribution, modal split, network assignment or the entropy models [21]). Particular attention has also been devoted to the description of residential areas and to demographic and industrial development on a regional scale, as well as to the causes and effects of migration phenomena. The reliability of the solutions of all the above problems is obviously strongly dependent upon the availability of a well structured information system: this is why the problem of geographical data collection and processing has also been thoroughly investigated.

Planning

Many regional planning problems have a discrete formulation such that the corresponding models are often integer (or mixed-integer) mathematical programs or combinatorial optimization problems which have to deal with particular configurations on a given graph. Among the most significant problems and models we can recall [22] the optimal partitioning of a given region into zones or districts, the location of plants and services, the determination of the shortest path between two nodes in a graph, the scheduling of a repetitive service, the determination of the optimal dimension and configuration of a transportation network, the selection among alternative public services, and so on. The applications [23] are just as numerous and interesting.

Forecasting

Forecasting techniques are often used in this field to predict exogenous variables. In the majority of cases the methods are not at all sophisticated (regression and time series analysis). Well documented applications refer to service demand forecasting, extrapolation of demographic data, evaluation of migration rates [24], and prediction of regional economic development.

Management

Problems of real-time management of urban and regional systems have only recently been considered. The most well-known example of such problems is traffic control in all its aspects: urban areas, access to freeways and recreation areas, daily adaptation of train schedules, real-time rerouteing of airplanes, 'dial a ride' transportation systems [25], and so on. Some other management problems of a completely different nature, like flat renting control in a metropolitan area, industrial and agricultural development through taxes and subsidies, and others have also been considered.

On the basis of the above analysis, we can conclude that the systems approach has allowed the synthesis of old and classical results on spatial distribution phenomena and has produced the possibility of tackling large-scale, and hence realistic, problems [26]. Significant contributions from other fields like geography, industrial management, and urban analysis have been lumped together in a unique conceptual body based on systems and decision theory (see, for example, the special issue of the *European Journal of Operations Research* on 'locational decision theory'

[27]). Simulation and planning models have been already developed and applied in a number of real cases. Research will probably be devoted in the near future to large-scale integrated studies related to real land use problems. Forecasting models have also been developed, but the most recent and sophisticated techniques have not yet been applied. On the contrary, research on management problems is still at the pioneering stage, particularly in the context of real applications. Nevertheless, qualified research programmes have already been defined and undertaken, like the project of the Organization for Economic Cooperation and Development on non-conventional transportation systems in non-densely populated areas.

Research activities on all modelling aspects of regional development and land use are currently in progress in many universities and research institutes. Among them, the International Institute for Applied Systems Analysis (Laxenburg, Austria) deserves special mention for its ambitious (five-year-long) research plan.

The most important theoretical contributions in this sector can be found in the journals *Operations Research, Networks*, and *Management Science*, while well documented application-oriented studies are often published in *Transportation Science, Journal of Regional Science, Town and Planning Review*, and *Urban Studies*.

AIR POLLUTION

Although air pollution has only recently become a popular subject of controversies and heated debate, it has been studied for a long time, with particular reference to transportation and the diffusion of pollutants in the atmosphere. The first very simple models, directly derived from the classical Fick's equation, have become sophisticated and have been subsequently extended within the framework of atmospheric physics [28, 29], and applied more recently to the location of power plants and the design of stacks or other effluent discharge facilities. During the last two decades public opinion and local authorities have often emphasized the need for reliable and clear studies on the consequences of air pollution. No doubt the ascertainment of such an urgent need has induced many a research institute to start research programmes on this matter, in particular on systems analysis methods for environmental impact assessment.

Simulation

Methods for the quantitative analysis of air pollution phenomena are well established. The most common of them deal with the description of the temporal evolution of the concentration field of inert substances in the atmosphere. These methods are based on distributed parameter models (Fick's equation), on Lagrangian models [30], or one more simple models like Gaussian, 'puff', and 'box' models which are obtained from the preceding ones by introducing suitable simplifying assumptions (stationarity and smothness of the solution). From a conceptual point of view these models are quite satisfactory, since they explicitely take into account the physical characteristics of the problem. Nevertheless, their application requires an enormous amount of data on the emission rates of the sources, the pollutant flux through the boundary of the region under consideration, and the local meteorology (at least wind direction and intensity, and the vertical profile of temperature). Moreover, most models are often difficult to solve, as indicated by the large amount of computational schemes proposed for the integration of the advection-diffusion equations (see, for example, [31]). Models of this type have been applied to a number of real cases (St. Louis, Nashville, Los Angeles, Venice and few others). The agreement with the observed data has been very satisfactory for stationary conditions and long-term averages. On the contrary, when non-stationary conditions are considered the performance of these models is often inadequate and the amount of data needed for tuning the model may become excessive. More recently, a few models have been proposed and used in order to describe the complex interactions taking place in the atmosphere between different compounds in the case of photochemical smog [30]. Moreover, the effects of air pollution on animals, vegetation and human health (for example, the dynamic relationships between carbon monoxide and haemoglobin, and correlations between sulphur dioxide and lung diseases), have also been investigated by means of simple interpretative models. Finally, it is worthwhile to mention that the problems of acid rain and long-distance transport of pollutants in the atmosphere has recently attracted the interest of a number of qualified research groups.

Planning

The simulation models described above have been used for a long time [32] to solve in a naïve way the trade-off problem between cost of the plant (e.g. height of the stack) and environmental damage (e.g. corrosion of surrounding structures). More recently, the problem of optimal location and design of the pollution sources has been formulated in terms of mathematical programming (linear, non-linear, mixed-integer), but this approach is not yet followed in practice by the decision-makers. These planning models need a detailed economic description of all the benefits and costs associated with a given plant and a quantitative evaluation of the corresponding environmental damages. Since this damage is difficult to express in monetary terms, the technique likely to be applied in the future is multiobjective mathematical programming.

Forecasting

Real-time forecast of ground-level concentrations of air pollutants is becoming a popular subject of research, especially in view of the success of a number of applications in urban areas (for example, Osaka, Nagoya, Tokyo, London, and Milan), where the forecast is used as an alarm in situations of forthcoming severe episodes. In the case of industrial areas, the forecast is also used in order to undertake quickly effective countermeasures like load reduction or shifting and fuel switching [33, 34]. In some urban areas (e.g. Osaka) the access of cars to the centre of the town may be forbidden when too-high a concentration of air pollutants is predicted for the following hours. The forecasting technique which has been used so far is time series analysis [35, 36], with the exception of a few studies where adaptive Kalman filtering has been applied in order to take into account the variability of the meteorology-dependent parameters [35, 37]. In some of the Japanese urban air pollution centres a technique based on geographical data analysis and pattern recognition has also been implemented.

Management

Although the applications (in particular in industrial areas) are relatively numerous, only a few cases have been documented in the technical literature. Nevertheless, this is an area of great interest and concern and a substantial research effort should be expected in the near future.

The above analysis confirms that the most relevant air pollution problems can be formally described by means of models. Unfortunately, the most classical models, namely those suggested by the real mechanisms of the phenomenon, are quite complex and often non-linear because of the chemical reactions occurring in the atmosphere. Thus, the progress of the whole sector seems to be strategically constrained by the possibility of developing approximate low-order models. Another crucial point for the succes of the systems approach in this sector is the need of effective techniques for identifying simple but significant meteorological indicators (categories) that might be used to reduce the amount of data necessary for simulation and validation.

Highly qualified research programmes are currently developed by a number of research institutions. Among them we can recall the National Environmental Research Center of EPA (Triangle Park, N.C., USA), the Chemical Engineering Department of the Californian Institute of Technology (Pasadena, Cal., USA), the Institute of Atmospheric Physics of the University of Berlin (West Berlin, Federal Republic of Germany) and the Main Geophysical Observatory of the University of Leningrad (Leningrad, USSR).

As far as publications are concerned, the best contributions dealing with the technical aspects of the problems can be found in the journals *Atmospheric Environment, Journal of Applied Meteorology, Journal of Air Pollution Control Association*, and the *Journal of Atmospheric Sciences*. For the economic aspects and the relationships with land use and industrial planning it might be useful to refer, among others, to the *Journal of Environmental Economics and Management*.

WATER POLLUTION

From the beginning of the century, and in particular in the last thirty years, the local authorities in charge of water pollution control have more and more frequently asked research institutions to develop tools for the quantitative evaluation of the economic and environmental impact of proposed river basin plans. This is why, for example, Streeter and Phelps proposed in 1925 a model for the determination of the dissolved oxygen concentration in the Ohio river. Since then, such a model has been extensively used both in Western and Eastern countries. Only in the last ten years research has been directed toward more complex methods and problems [38–42].

Simulation

The simulation of the dissolved oxygen in a stretch of river is certainly the problem which has received more attention to date, particularly in the case of biodegradable pollution [41,42]. More recently, eutrophication of lakes and estuaries has also been considered [43, 44], with due attention being given to the role played by hydraulic circulation. On the other hand, problems of pollution of marine systems [45], of salt intrusion in coastal aquifers, and of groundwater pollution [46] have been given little attention until now. Many of the above models and the techniques for tuning such models have been adapted recently to the description of the bacteria growth in wastewater treatment plants [47].

Planning

The most common planning problem is the design of a network of wastewater treatment plants in a region in which effluent and/or stream standards are not met. This problem has been modelled in various ways by different authors [41]. For example, there are models which allow one to determine the cheapest treatment network which satisfies a set of stream standards, models which determine among all possible networks at a given cost the one which minimizes the environmental impact, and others which allow one to solve the tradeoff between transportation and treatment costs. All these models have been used in real cases or tested with realistic data, and some of them have been programmed and are now available in standard form together with a user manual. Research on the consequences and impact of different environmental laws (standards, taxes, subsidies, etc.) is much less advanced, although the problem is obviously very important. Recent studies on this matters (see, for instance, [48]) make use of 'game theory' and the first results are definitively encouraging, at least on a conceptual level.

Forecasting

Water pollution forecasting models have been developed only rarely. Some studies on flow time and on propagation of pollutants in a river have been carried out with the aim of developing reliable alarm systems in case of sudden discharges in the upper parts of a river. Such studies are based on very sophisticated decomposition techniques [41], which are certainly at the frontier of the theory of optimal filtering and prediction.

Management

So far, optimal control theory has been used mainly for real-time operation of wastewater treatment plants [47]. Only a few theoretical contributions are available (see for instance [41]) for real-time management of rivers and lakes.

The contribution of systems analysis in this field is almost exclusively limited to simulation, but is, nevertheless, fundamental. In fact, 'trial and error' methods for tuning the models have been substituted by more effective automatic techniques derived from identification theory. Moreover, it is worthwhile to mention the discussion initiated in the early seventies on the possibility of interpreting the problem of the identification of the various components on the aquatic ecosystem (bacteria, protozoa, algae, etc.) as a 'state reconstruction problem'. This fact, which has already suggested a number of interesting studies, should be further investigated by means of non-linear filtering techniques. Finally, it is also important to point out that the 'systems approach' has allowed the development of a number of theoretical results, mainly related to the Streeter–Phelps model, which already constitute material for an advanced course on water quality modelling.

The most interesting research activities in this field are carried out by a number of very small research teams. On the other hand, applied research programmes characterized by extensive data collection and laboratory experiments are carried out by research centres such the Center for Environmental Research Information of EPA (Cincinnati, Ohio, USA), the Canadian Centre for Inland Waters (Burlington, Ontario, Canada), and the Institute of Hydrology (Wallingford, UK).

As for the technical literature, the journals which publish the best contributions on water pollution modelling are *Water Resources Research*, the *Journal of Environmental Engineering Division* of the ASCE, the *Journal of Environmental Economics and Management*, and the *Transactions on Systems, Man and Cybernetics* of the IEEE.

ECOLOGY AND RENEWABLE RESOURCES

As is well known, ecology deals with the study of the interactions and mechanisms which determine the dynamics and the spatial distribution of all living species (including man). When a species is exploited by man (for economic reasons or others) it is called a 'renewable resource' to stress the fact it is a resource capable of reproduction. The first descriptive ecological models were due to Lotka (1924) and Volterra (1926) who independently studied the dynamic interactions between predators and prey by means of a set of ordinary differential equations, while the first quantitative studies on the management of renewable resources (fisheries) were based on the theory of the sigmoid curve of Graham (1935). Since then, quantitative studies in ecology and renewable resources have been more and more frequent (for a survey, see [49]) and systems analysis techniques are now recognized by ecologists as essential tools.

Simulation

The description of ecosystems is certainly the problem which has attracted most investigation in quantitative ecology. The proposed models can be divided into two classes depending upon whether temporal or spatial variability is analysed. Correspondingly, one refers to 'population dynamics' in the former case, and to 'spatial heterogeneity' in the latter. A fairly balanced and detailed development of both problems can be found in the book by Pielou [50], while a more synthetic and original approach has been presented by MacArthur and Wilson [51]. Population dynamics is a well established field characterized by its own theory. Typical topics of such a field are the stability of complex ecosystems, and the theory of competition [52]. Quite important also is the estimation of the parameters characterizing the dynamics of the population on the basis of historical data (see, for example, [53]). Experiments on this matter have been carried out mainly on laboratory populations, but also on some natural populations. As far as the spatial patterns are concerned, less effort is devoted to the general principles than to the analysis of real data (e.g. by multivariate analysis). Typical problems are the classification and ordination of species, the determination of species-abundance relationships, and the definition of indicators of ecological diversity. These problems are particularly important in botany, where a classical textbook is the one by Orloci [54].

Planning

So far, the determination of long-term strategies for the rational exploitment of renewable resources has not been given particular attention. This is possibly due to the unpredictability of the future evolution of exploited ecosystems which makes the real-time management problem more meaningful than the planning problem. Exceptions can be found, of course, in agriculture and forestry [55], where long-term planning is of vital importance. In fisheries planning, the notion of maximum sustainable yield has been used for a long time but has been smoothly substituted by more realistic decision-making criteria based on detailed simulation and uncertainty analysis (for a review, see [56]). It is quite surprising that operations research techniques have not yet been extensively applied to aquaculture, where environmental fluctuations are rather limited and population dynamics is under control.

Forecasting

Forecasting models have been only rarely developed to predict the possible evolution of an ecosystem. Problems of this kind can be encountered when estimating the biological parameters of a population, or when developing short-term forecasts for a manager. However, although the question 'how will an ecosystem evolve?' is of undoubted value, the use of effective techniques such as Kalman filtering and time series analysis has received only little attention [57]. A possible justification for this stems from the fact that these techniques make reference to linear models which only rarely can be reasonably used in ecology.

Management

Research on ecosystems management is today a very popular subject, and the contemporary application of the basic notions of biology and economics within the framework of systems analysis seems to be very promising. The book by Watt [58] devoted to fishing and hunting management problems, to wildlife conservation, and to pest control was certainly the first sound treatise on this line. Research making explicit use of optimal control theory is developed and advanced. Extremely interesting, from this point of view, is the book by Clark on *Mathematical Bioeconomics* [59] which is mainly devoted to the optimal exploitation of a single species. Recently, more complex and realistic problems like the exploitation of multispecies ecosystems, conflict resolution, and non-cooperative situations have also been explored to a certain extent (see for instance [60], [61]).

As for the development of this sector, we can summarize the above analysis by saying that a relevant research effort will certainly be devoted during the next ten years to the following topics: influence of spatial heterogeneity on ecosystems dynamics, reduction of complex simulation models to compressed low-order models, application of operations research to agriculture and aquaculture, development of non-linear estimation and forecasting techniques, application of multiobjective mathematical programming and game theory to renewable resources management.

Because of the variety of problems, qualified research institutes in this sector are numerous. Among them we might list the Institute of Animal Resource Ecology of the University of British Columbia (Vancouver, B.C., Canada), the Department of Biology of Stanford University (Stanford, California, USA), the School of Fisheries of the University of Washington (Seattle, Washington, USA), the Biology Department of Princeton University (Princeton, New Jersey, USA), the Section of Ecology and Systematics of Cornell

University (Ithaca, New York, USA), and the Department of Mathematics of the University of British Columbia (Vancouver, B.C., Canada).

As far as the technical literature is concerned, it is worthwhile to note that papers on the topics discussed above are published not only by the journals explicitely devoted to mathematical modelling in biology and ecology (like *Mathematical Biosciences* and *Ecological Modelling*), but also by journals of traditional experimental and naturalist character (like *Ecology* and *The American Naturalist*), or even by journals devoted to specific fields (like the *Canadian Journal of Fisheries and Aquatic Science*). Moreover, systems-oriented journals (like the *Journal of Optimization Theory and Applications*) are also nowadays publishing, with more and more interest, contributions in the field of ecological modelling.

A COMPARATIVE ANALYSIS

The material presented in the preceding sections deserves some further discussion before we reach definite conclusions. Hence, we will make reference to the matrix of Figure 1, where the rows are the five sectors I have analysed (water resources, land use, air pollution, water pollution, and ecology) and the columns correspond to the problems (simulation, planning, forecast, and management) I have identified. In the matrix the three tones (black, grey, and white) are used to classify the status of research, teaching and consulting activities as follows.

First class (black): basic research already developed, numerous applications and case studies well documented, software largely available in standard form.

Second class (grey): basic research still in progress, some applications to real or realistic cases recently published, software available only for a few specific problems.

Third class (white): basic research at an early stage, applications to real cases not yet documented, software not yet developed.

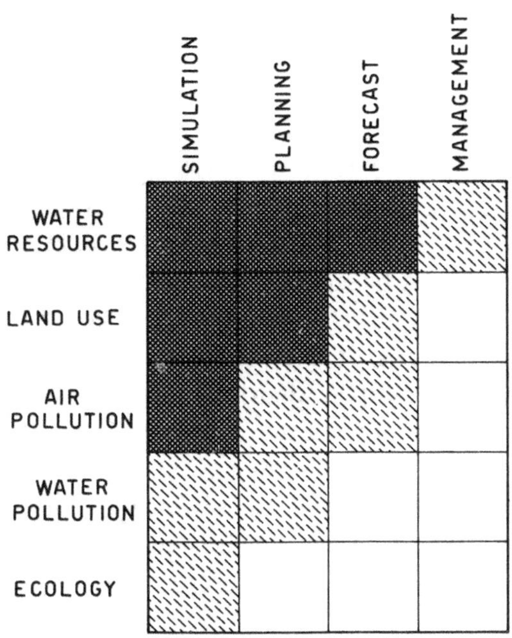

Fig. 1. Classes of high, medium and low development.

It is worthwhile noticing that the classification is not based upon the total amount of publications one can find in the literature, but upon the comparison of the current state of the art with the potential of each sector. Thus, for example, the publications devoted to water resource management are about the same, if not more, than those concerning simulation models of air pollution. Nevertheless, water management belongs to the second class (grey), while air pollution simulation is in the first class (black). This means that research on the first topic is still in progress, while the second subject has already largely developed. Obviously, the classification presented in Figure 1 is rather subjective, although it is based on the elements discussed in the previous sections. Nevertheless, a number of independent remarks can be made to confirm this classification.

First of all, the figure shows that in each sector (row) one goes from a fairly developed state (black or grey) to a developed one (grey or white). This structural result is, in a sense, justifiable *a priori* since planning problems can be solved only after satisfactory simulation models have been developed, and real-time management often requires the use of reliable forecasts. Moreover, planning problems have historically received more attention than management problems because of their major economic and social impact.

A systematic analysis of the columns also shows that the first sectors are more developed than the others. This is also easily understandable if one notices that water pollution studies require a detailed knowledge of hydrology (first row), that air pollution is often easier to understand than water pollution because in the second case the dynamics of the aquatic ecosystem plays an essential role, and that ecology, possibly because of the complexity of its phenomena, has still some difficulty in becoming a 'hard' science.

A check on the validity of the classification of Figure 1 can be made by carrying out some simple statistics on the most significant journals of each sector. Thus, for example, in the last five years more than half of the papers published in the journal *Water Resources Research* are devoted to modelling-oriented studies, in the following proportions:

simulation	27 percent
planning	14 percent
forecasting	8 percent
management	5 percent

while twenty years ago contributions on forecast and management were almost non-existent. Statistics carried out on the journal

Ecological Modelling since its first issue in 1975 show a dominant interest in simulation studies (more than 90 percent of the papers) and point out that interest in planning and forecasting problems is marginal. Similar results confirming the classification of Figure 1, have been obtained by analysing the journals mentioned in the sections above.

A systematic analysis of the graduate and postgraduate courses on environmental systems analysis is also of interest. The first courses were taught in USA in the late 1960s. Since then, a great variety of academic curricula largely based on systems analysis have been created in schools of engineering, architecture, and applied sciences. These curricula are based on numerous courses which specifically deal with the sectors and problems appearing in Figure 1. In particular, in a large number of engineering schools the sector of water resources is well developed and masters-level courses on simulation, planning, forecast and management of water resources are available. Teaching activities in the field of land use are also rather well developed: for example, many universities offer modelling-oriented courses on urban systems analysis and on land use planning. Much less is available, on the other hand, on the last three sectors where not a single course exists devoted to one problem of the four mentioned (simulation, planning, forecast and management).

Finally, an analysis of the nature and frequency of symposia and conferences devoted to this field and held during the last ten years is interesting. On one side we have a number of congresses devoted to environmental systems analysis as a whole; and among them we can list the following

IFAC–UNESCO Workshop on 'Systems Analysis and Modelling Approach in Environmental Systems', 1973.

CAS International Seminar on 'Systems Approach to Environmental Problems', 1975.

IFIP Working Conference on 'Environmental Modelling', 1975.

IFIP Working Conference on 'Modelling and Simulation of Land, Air, and Water Resources Systems', 1977.

IFAC Symposium on 'Environmental Systems Planning, Design and Control', 1977.

IFIP Working Conference on 'Simulation and the Environment', 1979.

IFIP Working Conference on 'Environmental Systems Analysis and Management', 1981.

On the other hand, we have a large number of very specialized meetings devoted to very particular but important topics (like, for example, the Pittsburgh Conference on the applications of the Kalman filtering techniques to hydrology [14], the Waterloo Symposium on real-time management of hydrosystems [17], the Vienna Workshop on air pollution modelling [35], and the Christchurch Workshop on renewable resources management [62]). The tendency towards meeting specialization is certainly detectable, in particular for the first two sectors, and this again confirms the validity of the classification.

Now that Figure 1 has been somehow 'validated', we can derive from it some indications about the most likely future evolution of the demand of qualified professionals in all different sectors. For this purpose, let me distinguish the activities in

- research
- teaching
- consulting

which certainly represent the three major branches of employment for environmental systems analysis.

Research activities are mainly carried out at the university and deal with the assessment of general theories and the development of new techniques for the analysis of environmental systems. These activities are particularly significant in those sectors where there is more need of a conceptual framework and systems analysis has not yet been applied for solving real problems. These studies are the subject of numerous Masters and Ph.D. theses, where the methodological aspects of the problems are particularly emphasized. Sometimes these theses make reference to real problems which are nevertheless often oversimplified. Still a great part of these problems falls in the first class of Figure 1. Nevertheless, these topics are somehow considered too classical by the most active universities, which indeed are more concerned with research programmes in the second and third class of Figure 1. In conclusion, one can reasonably predict that in the next ten years the research effort will propagate along the diagonal of our matrix, as shown in Figure 2.

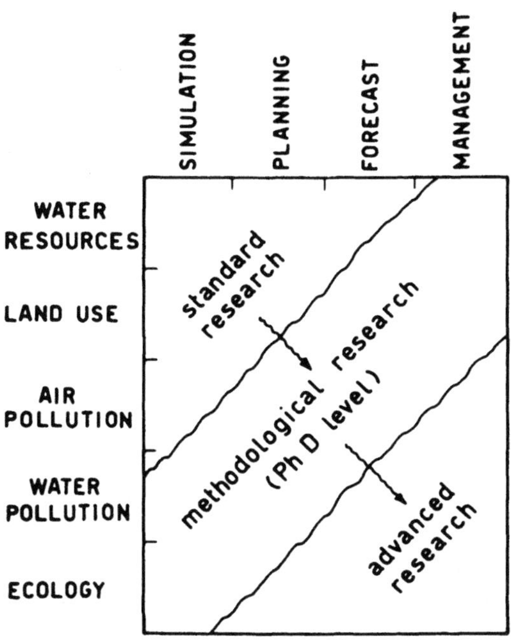

Fig. 2. Research activities.

Teaching activities at graduate and postgraduate level are obviously strongly correlated with research activities. The three classes shown in Figure 1 roughly correspond to 'graduate' courses, and research seminars as shown in Figure 3.

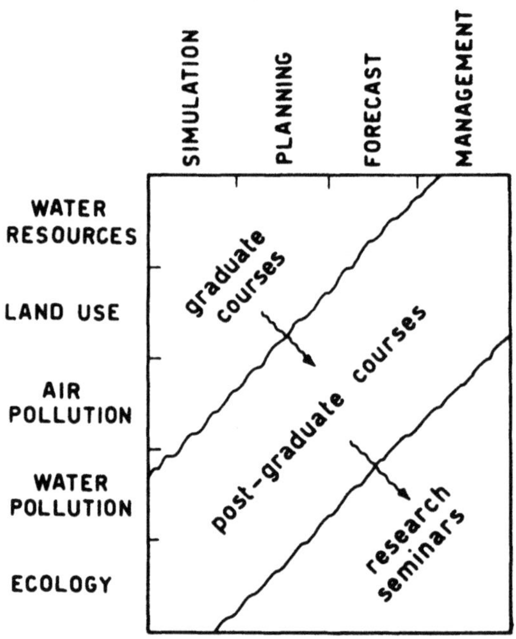

Fig. 3. Teaching activities.

The applications of environmental systems analysis are usually carried out by the research centres of the state and local authorities in charge of land use and pollution control, or by consulting groups. Obviously, more studies are in progress in those fields where the 'system approach' has already proved its power, and validated software is available. Such studies are standard for consulting groups qualified in systems analysis, and, on the other hand, such a qualification is more and more frequently and explicitely required to avoid the use and abuse of non quantitative techniques of analysis. The activity of the above research centres and consulting groups is only rarely concerned with the second class of problems shown in Figure 1 and still far from the problems of the third class. We can therefore visualize this situation on the matrix of Figure 4.

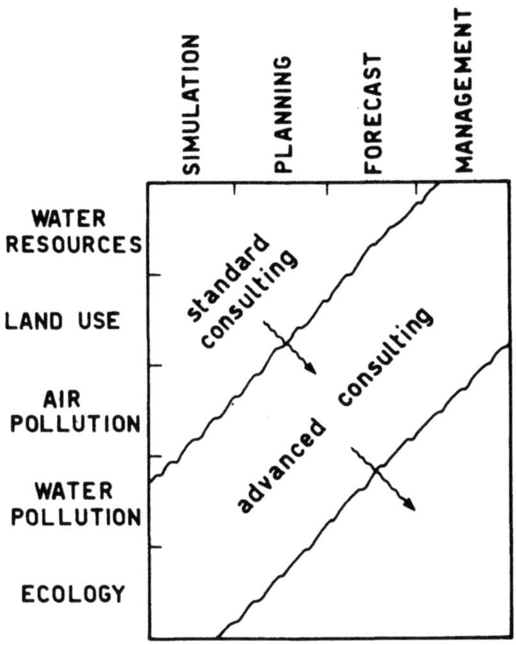

Fig. 4. Consulting activities.

In conclusion, comparative analysis of the different sectors and problems of environmental systems analysis, allows us to predict that during the next ten years the three major sources of employment, namely consulting, teaching, and research, will be concentrated, respectively, on the first, second, and third class of problems shown in Figure 1. The situation will, of course, evolve smoothly and the interest should gradually shift toward less developed classes.

CONCLUSIONS

The analysis of some sectors of environmental sciences and of their characteristic problems has allowed us to group all research topics in three different classes, respectively of high, medium and low development. The first class contains simulation, planning, and forecasting of water resources systems, analysis and planning of urban and regional systems, and the simulation of air pollution phenomena. Such problems can be solved by means of well established models which are taught in Masters-level courses and regularly used in consulting. This area should attract those who are interested in developing consulting firms which may become productive within a very short timescale. On the other hand, long-term planning of such activities should be more concerned with the second class of problems, namely water resources management, forecasting techniques for urban and regional systems, air pollution planning and forecasting, simulation and planning of water pollution, and simulation of 'productive' ecosystems (forestries, fisheries and agriculture). This class of problems should also attract those who are interested in an academic career, since a great number of courses dealing with these problems will certainly start within the next few years. Finally, the third class of problems, which is more concerned with real-time forecasting and management in 'unexplored' sectors such as water pollution and ecology, should be given particular attention in planning the long-term development of research institutions.

ACKNOWLEDGEMENTS

The author wishes to thank his colleagues Drs. A. Colorni, G. Finzi, G. Fronza, M. Gatto, G. Guariso, E. Laniado, and R. Soncini-Sessa for their helpful suggestions and comments.

REFERENCES

[1] S. Rinaldi, 'Environmental System Engineering at the Politecnico di Milano', IIASA Workshop on Energy, Economics, Technology and Environment (in Italian) (Rome, Italy, February, 1981).

[2] GRAGSA, *Environmental Systems Analysis and Management* (in Italian) (CLUP, Milano, Italy, 1979).

[3] S. Rinaldi, 'Comments on the UNESCO Workshop on the Application of Systems Analysis to Environmental Management and Engineering' (Udine, Italy, December 1978, UNESCO report, Paris, France, 1979).

[4] S. Rinaldi (ed.), *Environmental Systems Analysis and Management*, Proceedings of IFIP Working Conference (North-Holland, 1982).

[5] A. Maas *et al.*, *Design of Water Resource Systems* (Harvard University Press, Cambridge, Mass., 1962).

[6] The Hydrologic Engineering Center, 'Publications Catalog' (HEC, Davis, California, October 1981).

[7] D. B. Simons, *et al.*, 'Sedimentation Study of the Yazoo River Basin' (General Report, Department of Civil Engineering, Colorado State University, Fort Collins, Colorado, 1978).

[8] J. Doorenbos and A. H. Kassam, 'Yield Response to Water', FAO Irrigation and Drainage, Paper n. 33, FAO (Rome, Italy, 1979).

[9] Y. Y. Haimes, *Hierarchical Analyses of Water Resources Systems* (McGraw-Hill, New York, 1977).

[10] D. P. Loucks, J. R. Stedinger and D. A. Haith, *Water Resource Systems Planning and Analysis* (Prentice-Hall, N.J., 1981).

[11] D. Marks, 'Models in Water Resources', in *A Guide to Models in Governmental Planning and Operation* (EPA publication, Washington D.C., 1974).

[12] R. G. Cummings, *Interbasin Water Transfer: A Case Study in Mexico*, Resources for the Future (Johns Hopkins University Press, Baltimore, Md., 1974).

[13] M. Fiering, *Steamflow Synthesis* (Harvard University Press, Cambridge, Mass., 1967).

[14] C. L. Chiu (ed.), *Applications of Kalman Filter to Hydrology, Hydraulics, and Water Resources* (University of Pittsburgh Press, Pittsburgh, 1978).

[15] C. W. Howe, and F. P. Linaweaver, 'The Impact of Price on Residential Water Demand and Its Relation to System Design and Price Structure', *Water Resources Research* **3** (1) (1967), 13–32.

[16] J. Kindler (ed.), *Forecasting Water Demand* (Pergamon Press – IIASA, Oxford, 1983).

[17] T. E. Unny and E. A. McBean (eds.), *Real-Time Operation of Hydrosystems*, Proceedings of an International Symposium held at the University of Waterloo, June 1981 (Water Resources Publications, Fort Collins, Colorado, 1982).
[18] W. Isard, *Methods of Regional Analysis* (John Wiley, New York., 1960).
[19] W. Alonso, *Location and Land Use* (Harvard University Press, Harvard, Mass., 1965).
[20] I. S. Lowry, *A Model of Metropolis* (RAND Corporation, Santa Monica, Cal., 1964).
[21] A. G. Wilson, *Entropy in Urban and Regional Modelling* (Pion, London, 1970).
[22] E. J. Beltrami, *Models for Public Systems Analysis* (Academic Press, New York, 1977).
[23] EPA, *A Guide to Models in Governmental Planning and Operations* (U.S. Environmental Protection Agency, Washington D.C., 1974).
[24] L. J. Castro and A. Rogers, 'Migration Age Patterns: Measurement and Analysis (Part 1); Cause-specific Profiles (Part 2) (International Institute for Applied Systems Analysis, Laxenburg, Austria, 1979).
[25] R. F. Kirby *et al.*, *Para-transit: Neglected Options for Urban Mobility* (The Urban Institute, Washington, D.C., 1974).
[26] J. B. McLoughlin, *Urban and Regional Planning: A System Approach* (Faber and Faber, London, 1969).
[27] Various authors, 'Locational Decisions', *European Journal of Operations Research*, Special Issue, **6** (2) (1981).
[28] O. G. Sutton, *Micrometeorology* (McGraw-Hill, New York, 1953).
[29] F. Pasquill, *Atmospheric Diffusion* (Van Nostrand, New York, 1962).
[30] J. H. Seinfeld, *Air Pollution Physical and Chemical Fundamentals* (McGraw-Hill, New York, 1975).
[31] R. D. Richtmeyer and K. W. Morton, *Difference Metbods for Initial Value Problems* (Interscience, New York, 1967).
[32] J. M. Guldman and D. Shefer, *Industrial Location and Air Quality Control* (J. Wiley, New York, 1980).
[33] P. Melli, P. Bolzern, G. Fronza, and A. Spirito, 'Real-time Control of Sulphur Dioxide Emissions from an Industrial Area', *Atmospheric Environment* **15** (1981), 653–66.
[34] P. Bolzern and G. Fronza, 'Cost-effectiveness Analysis of Real-time Control of SO_2 Emission from a Power Plant', *Journal Environmental Management* **14** (1982), 253–63.
[35] G. Fronza and P. Melli (eds.), *Mathematical Models for Planning and Control of Air Quality* (Pergamon Press – IIASA, Oxford, 1982).
[36] G. Finzi and G. Tebaldi, 'A Mathematical Model for Air Pollution Forecast and Alarm in an Urban Area', *Atmospheric Environment* **16** (1982), 2055–9.

[37] S. G. Bankoff and E. L. Hanzevack, 'The Adaptive Filtering Transport Model for Prediction and Control of Pollutant Concentration in an Urban Airshed', *Atmospheric Environment* **9** (1975), 793–808.
[38] R. Dorfman, H. D. Jacoby, and H. A. Thomas, *Models for Managing Regional Water Quality* (Harvard University Press, Cambridge, Mass., 1972).
[39] R. V. Thomann, *Systems Analysis and Water Quality Management* (McGraw-Hill, New York, 1972).
[40] R. M. Keinath and P. M. Wanielista, *Mathematical Modelling for Water Pollution Control Processes* (Ann Arbor Science, Ann Arbor, Mich., 1975).
[41] S. Rinaldi, R. Soncini-Sessa, H. Stehfest, and H. Tamura, *Modelling and Control of River Quality* (McGraw-Hill, New York, 1979).
[42] A. K. Biswas (ed.), *Models for Water Quality Management* (McGraw-Hill, New York, 1981).
[43] E. J. Middlebrooks, D. E. Falkenborg, and T. E. Maloncy, *Modelling the Eutrophication Processes* (Ann Arbor Science, Ann Arbor, 1974).
[44] G. T. Orlob, 'Estuarial Models', in *Systems Approach to Water Management*, A. K. Biswas (ed.) (McGraw-Hill, New York, 1976).
[45] J. Nihoul (ed.), *Modelling of Marine Systems* (Elsevier, Amsterdam, 1975).
[46] J. J. Fried, *Groundwater Pollution* (Elsevier, Amsterdam, 1975).
[47] J. F. Andrews, 'Dynamic Models and Control Strategies for Wastewater Treatment Plants: An Overview', IFAC Symposijm on Environmental Systems Planning, Design, and Control (Kyoto, Japan, 1977).
[48] S. Rinaldi, R. Soncini-Sessa and A. B. Whinston, 'Stable Taxation Schemes in Regional Environmental Management', *Journal of Environmental Economics and Management* **6** (1979), 29–50.
[49] S. Rinaldi, 'Modelling and Control of Renewable Resources', IFAC Symposium on Environmental Systems Planning, Design, and Control (Kyoto, Japan, 1977), published in *Ricerche di Automatica* **9** (1) (1978), 15–33.
[50] E. C. Pielou, *Mathematical Ecology* (J. Wiley, New York, 1977).
[51] R. H. MacArthur and E. O. Wilson, *The Theory of Island Biogeography*, Monographs in Population Biology (Princeton University Press, Princeton, 1967).
[52] R. H. MacArthur, 'Species Packing, or What Competition Minimizes', *Proc. Natl. Acad. Sci. USA* **64** (1969), 1369–75.
[53] P. H. Leslie, 'An Analysis of the Data for Some Experiments Carried Out by Gause with Populations of the Protozoa *Paramecium aurelia* and *Paramecium caudatum*', *Biometrika* **44** (1957), 314–27.
[54] L. Orloci, *Multivariate Analysis in Vegetation Research* (Junk, The Hague, 1975).

[55] G. R. Gregory, *Forest Resource Economics* (Ronald Press, New York, 1972).
[56] K. B. Haley (ed.), *Applied Operations Research in Fishing* (Plenum Press, New York, 1981).
[57] R. W. Poole, 'Empirical Multivariate Autoregressive Equation Predictors of the Fluctuations of Interacting Species', *Math. Biosci.* **28** (1976), 81–97.
[58] K. E. F. Watt, *Ecology and Resource Management* (McGraw-Hill, New York, 1968).
[59] C. W. Clark, *Mathematical Bioeconomics* (J. Wiley, New York, 1976).
[60] M. Gatto and S. Rinaldi (eds.), 'Ecology, Renewable Resources and Optimal Control', *Ecological Modelling*, Special Issue, **14** (3/4), (1982).
[61] C. J. Walters and R. Hilborn, 'Adaptive Control of Fishing Systems', *J. Fish. Res. Board Can.* **33** (1976), 145–59.
[62] T. L. Vincent and J. M. Skowronski (eds.), *Renewable Resource Management*, Lecture Notes in Biomathematics (Springer Verlag, Berlin, 1981) (Proceedings of a Workshop on Control Theory Applied to Renewable Resource Management and Ecology, Christchurch, New Zealand, 1980).

Centro Teoria dei Sistemi, CNR, Politecnico di Milano, Italy

M. Aoki

Systems Sciences and the Modelling and Analysis of Dynamic Economic Phenomena

INTRODUCTION

Predicting future trends is hazardous under the best of circumstances. Attempting to forecast future roles of control and systems sciences in economic modelling and analysis is fraught with danger because the forecasts must necessarily cover a wide range of areas. In an effort to lessen the chances of gross misforecasts we report on recent and current contributions of system and control theoretic approaches to dynamic analysis in economics for indications and suggestions of future roles in some selected areas.

Broadly speaking, systems science can contribute in two related ways: by providing an alternative conceptual framework for dynamic analysis in areas where static or comparative static analysis still is a dominant mode of analysis and by developing new tools to facilitate dynamic or comparative dynamic analysis.

Just as the understanding of linear and nonlinear programming methods has popularized modelling and analysis of many problems as static constrained optimization problems, so dynamic programming, maximum principle, control and system theory can and have stimulated the formulation of many decision problems such as dynamic (constrained) optimization, i.e. intertemporal decision problems by individuals, firms, governments or policy makers. What used to be amenable only to static analysis may now be analysed as intertemporal problems incorporating many of the linkages which had to be ignored in earlier analysis. Nowhere is this reformulation of problems as dynamic ones more important than in policy models in the national and international settings. More and more problems will be formulated as intertemporal decisions problems as better analytical tools and computational algorithms become available.

Recent work on open-economy macroeconomics or investment problems by firms, and the use of overlapping generation models to pose questions on optimal consumption and investments are but a few examples which presage wider future uses of dynamic problem formulations. Increasingly, numerically stable solution algorithms and more precise control and prediction algorithms for dynamic systems are becoming available, which make solution of intertemporal optimization problems easier and faster and help the trend towards more extensive use of dynamic problem formulation. See [9] for a general account of state-space representation in macroeconomics and microeconomics. Applications in open-economy macroeconomics are found in [6]. Examples of maximum principle in intertemporal optimization may be found in [1], [10] or [22].

MODEL SPECIFICATION AND ANALYSIS IN ECONOMICS

Models mean different things to different people. They play diverse roles in disciplines such as mathematics, physics, medicine, and economics on which we focus. Very often the models' role as predictors of the outcome of experiments or future events are emphasized to the exclusion of all other functions. No doubt predictive capability is important in some models, but this ability is not necessarily their only or most important attribute. For example Kac [18] discusses a model's role in physics as means to illustrate subtle points in theory, even though models may not be realistic enough to serve as predictors of experimental outcomes.

A similar distinction can also be made for economic models. Although econometric models may immediately come to mind at the mention of models in economics, they are not the only kind. Different models are designed to meet different requirements to varying degrees. Predictive ability is important in econometric models but not in others such as those designed to study policy effects, or in some types of models in microeconomics (see [17] for an example of such a view; also [11]). Simplicity, elegance, fruitfulness, consistency or mere ability to help elucidate and organize complex phenomena, such as effects of policy instruments under alternative and conditional ways, are some other attributes all of which need be considered in judging a model. Realism or higher accuracy of prediction is, then, only one desirable attribute out of many. No model possesses these desirable attributes equally.

Then there is the question of how to represent models. Systems science has long popularized the use of state-space or Markovian representation, i.e. first-order vector differential or difference equation description of dynamic systems. This mode of representing economic models is an alternative to other representations such as a system of lag operators, and has some advantages that are just beginning to be appreciated by some economists and econometricians (see [12] or [16] for example). We describe modelling of dynamic economic phenomena by state-space or Markovian representation by selecting two out of many potential application areas to illustrate that state-space form may be employed to advantage in economic analysis. The first is modelling and prediction of time series, and the second is the broad area covering macroeconomic analysis of interdependence and interactions of national economies and national policy instruments.

These two areas represent two different types of model used in economics. The first describes models of time series which emphasize predictive capabilities. Time series occur in many contexts in economics as well as in many other disciplines. Here, accurate prediction is most important. Thus any method of improving modelling of time series is certainly of great interest. The second emphasizes organizational advantages afforded by models and the sensitivity analysis of model properties with respect to assumptions. System science has contributed by developing 'realization theory' which constructs approximate models from data on input and output sequences or sequences of covariance matrices. Improved techniques and new or refined concepts related to dynamic systems permit the examination of dynamic phenomena which are ignored in traditional economic analysis, not because they are unimportant but rather because they are 'too messy' to handle by traditional techniques. The second area thus discusses dynamic modelling contributions to economics not necessarily by providing more precise predictions but rather by providing better insights into complex interdependence between economic units. See [3], [4], [5] for examples.

Modelling and Analysis of Time Series

Traditionally, statisticians, econometricians, and probabilities employ autogressive (AR) models, moving average (MA) models or their combination autoregressive-moving average (ARMA) models or its variant (ARIMA) to represent and analyse time series.[1] We describe here an alternative approach contributed mainly by control and system engineers which supplements the approaches employed by econometricians. This alternative way of modelling time series puts dynamic and interacting subunits into a collective whole to describe and analyse time evolution of the 'state' of the whole system in a particularly easy and well-organized manner. State space methods allow numerically convenient and insightful analysis of data by its alternative organization, as we later illustrate.[2]

System characteristics or environments changing with time usually require time-domain characterization. State space representation is a time-domain representation of dynamics as contrasted with frequency domain or transform method of dynamic system representation. Some economic concepts such as multipliers are inherently time domain notions and can easily be related to impulse response function characterization of system responses in the time domain rather than to transfer function description which is a notation in transform domain.

When system characteristics are time-invariant, we have a choice of alternative system descriptions in time domain or transfer domain. For example, given a stationary time series, it can be described by a state space model or a transfer function model, i.e. ARMA-type description. The relative merits of these alternative representations depend on many factors and are not clear-cut. If ARMA models are naturally given, then they may be employed 'as is' without transforming them into state space form. Even when this conversion is made, ARMA models generate state space models of canonical forms, i.e. the matrices associated with dynamics have certain prescribed patterns of 0 and 1's in addition to at most $2n$ free parameters where n is the highest order of lags of the ARMA models. This special structure means that the original ARMA model representation is more convenient for some purposes. When sequences of output covariances or impulse responses are given, these canonical state space representations of ARMA models are highly sensitive to numerical errors. Construction of approximate models seems easier in state space form, especially in view of recent results on Hankel matrix approximation [19]. These will be discussed later in this article.

We first establish the equivalence of these two modes of time series representation, i.e. we show that models in one representative can be converted into the other and vice versa if numerical errors are assumed away in the process.

Equivalence

AR models. Time series models can be put into state space form in several ways. It is probably most convenient to start with AR models.

We use the letter L to denote a backward shift, i.e. $Ly_t = y_{t-1}$. Then an AR model $y_t - a_1 y_{t-1} - a_2 y_{t-2} - \cdots - a_n y_{t-n} = u_t$ can be put as

$$(1) \qquad \phi(L) y_t = u_t ,$$

where

$$\phi(L) = 1 + \alpha_1 L + \alpha_2 L^2 + \cdots + \alpha_n L^n .$$

Here y's and u's are taken to be scalar-valued.[3] Introduce n auxiliary variables by setting

$$\begin{aligned} x_1(t) &= y_{t-n+1}, \\ x_2(t) &= y_{t-n+2}, \\ &\vdots \\ x_{n-1}(t) &= y_{t-1}, \\ x_n(t) &= y_t - u_t. \end{aligned}$$

Advancing time by one, and by the above definitions we can write

$$\begin{aligned} x_1(t+1) &= x_2(t), \\ x_2(t+1) &= x_3(t), \\ &\vdots \\ x_n(t+1) &= y_{t+1} - u_{t+1}, \\ &= -\alpha_1 y_t - \cdots - \alpha_n y_{t-n+1}, \\ &= -\alpha_1 x_1(t) + \cdots - \alpha_n x_n(t) - \alpha_1 u_t. \end{aligned}$$

Define a state vector by these variables as $x(t) = (x_1(t), \ldots, x_n(t))'$. Then $x(t)$ evolves with time according to a first-order difference equation, called state transition equation

(2) $\quad x(t+1) = Ax(t) + bu_t,$

where

$$A = \begin{bmatrix} 0 & 1 & 0 & \ldots & \ldots & 0 \\ 0 & 0 & 1 & \ldots & \ldots & 0 \\ 0 & \ldots & \ldots & \ldots & 0 & 1 \\ -\alpha_n & \ldots & \ldots & \ldots & \ldots & -\alpha_1 \end{bmatrix}, \qquad b = \begin{bmatrix} 0 \\ 0 \\ -\alpha_1 \end{bmatrix},$$

and y_t is related to $x(t)$ by an algebric relation, called an observation or output equation

(3) $\quad y_t = (0 \ \ldots \ldots \ldots \ 0 \quad 1)x(t) + u_t.$

The pair of Equations (2) and (3) constitutes a state space representation of (1).

ARMA models. Next consider a time series model described by

$$y_t - \alpha_1 y_{t-1} - \cdots - \alpha_n y_{t-n} = \beta_0 u_t + \beta_1 u_{t-1} + \cdots + \beta_{n-1} u_{t-n+1}.$$

It can be put into a form analogous to (1),

(4) $\quad \phi(L)y_t = \psi(L)u_t,$

where $\phi(L)$ is as defined above and we define

$$\psi(L) = \beta_0 + \beta_1 L + \cdots + \beta_n L^n.$$

Some of the β's may be zero. Its state space representation can be obtained in two steps. First define v_t by

(5) $\quad \phi(L)v_t = u_t.$

Then (4) can be written as

(6) $\quad y_t = \psi(L)v_t.$

The sequence $\{v_t\}$ generated by (5) has the same form as (1), hence can be put into state space form

$$(7)\qquad \begin{aligned} &z(t+1) = Az(t) + bu_t, \\ &\text{and} \\ &v_t = (0 \ \dots\dots\dots \ 0 \quad 1)x(t) + u_t, \end{aligned}$$

by introducing the vector $z(t) = (z(t), \dots, z_n(t))'$ where $z_1(t) = v_{t-n+1}, \dots, z_{n-1}(t) = v_{t-1}$ and $z_n(t) = v_t - u_t$. The state vector $z(t)$ is next related to y_t of (6) by

$$(8)\qquad \begin{aligned} y_t &= \beta_0 v_t + \beta_1 v_{t-1} + \dots + \beta_{n-1} v_{t-n+1} \\ &= \beta_0 (z_n(t) + u_t) + \beta_1 z_{n-1}(t) + \dots + \beta_{n-1} z_1(t) \\ &= c'z(t) + \beta_0 u_t, \end{aligned}$$

where

$$c' = (\beta_{n-1}, \dots, \beta_1, \beta_0).$$

Collecting (7) and (8) together, a state space representation of the ARMA model (4) becomes

$$(9)\qquad \begin{aligned} z(t+1) &= Az(t) + bu_t, \\ y_t &= c'z(t) = \beta_0 u_t. \end{aligned}$$

There are other ways of putting ARMA models into state space form. We discuss two such representations: controllable and observable.

For example, given

$$y_t = \alpha_1 y_{t-1} + \dots + \alpha_n y_{t-n} + \beta_0 y_t + \dots + \beta_n u_{t-n},$$

let

$$x_1(t) = y_t - \beta_0 u_t.$$

Then we can write the above as $x_1(t) = L(\alpha_1 y_t - \beta_1 u_t) + \cdots + L^n(\alpha_n y_t - \beta_n u_t)$, or

$$L^{-1}x_1(t) = x_1(t+1) = \alpha_1 y_t - \beta_1 u_t + x_2(t),$$

where we introduce a new variable by

$$x_2(t) = L(\alpha_2 y_t - \beta_2 u_t) + \cdots + L^{n-1}(\alpha_n y_t - \beta_n u_t).$$

Eliminating y_t by the first equation, the above becomes

$$x_1(t+1) = \alpha_1 x_1(t) + (\alpha_1 \beta_0 - \beta_1)u_t + x_2(t).$$

Rewrite the definitional equation for $x_2(t)$ as

$$\begin{aligned} L^{-1}x_2(t) = x_2(t+1) &= \alpha_2 y_t - \beta_2 u_t + x_3(t) \\ &= \alpha_2 x_1(t) + (\alpha_2 \beta_0 - \beta_2)u_t + x_3(t), \end{aligned}$$

where we introduce a new variable by

$$x_3(t) = L(\alpha_3 y_t - \beta_3 u_t) + \cdots + L^{n-2}(\alpha_n y_t - \beta_n y_t).$$

Continue in this way until we reach

$$x_n(t) = L(\alpha_n y_t - \beta_n u_t)$$

or

$$x_n(t+1) = \alpha_n x_1(t) + (\alpha_n \beta_0 - \beta_n)u_t.$$

Collecting $x_1(t+1), \ldots, x_n(t+1)$ as a column vector, we define it as the state vector. Then it obeys the state transition equation

$$(9') \qquad z(t+1) = \begin{bmatrix} x_1(t+1) \\ \vdots \\ x_1(t+1) \end{bmatrix} = Az(t) + bu(t),$$

and the output equation becomes $y_t = (1 \quad 0 \quad \dots\dots\dots \quad 0)z(t) + \beta_0 u_t$, where

$$A = \begin{bmatrix} \alpha_1 & 1 & 0 & \dots\dots\dots & 0 \\ \alpha_2 & 0 & 1 & 0 \quad \dots\dots & 0 \\ \alpha_{n-1} & \dots & \dots & \dots\dots\dots & 1 \\ \alpha_n & 0 & \dots & \dots\dots\dots & 0 \end{bmatrix} \quad \text{and} \quad b = \begin{bmatrix} \alpha_1\beta_0 - \beta_1 \\ \alpha_n\beta_0 - \beta_n \end{bmatrix}.$$

Compare the alternative state space representation of (9) and (9′). The representation (9) has a simple b vector with a complicated c vector, while (9′) has a complicated b vector and a simple c vector. The form (9) is called a controllable representation because the n-vectors $b, Ab, \dots, A^{n-1}b$ are easily seen to be linearly independent, because they have the structure

$$b = \begin{bmatrix} 0 \\ \cdot \\ \cdot \\ \cdot \\ 0 \\ x \end{bmatrix}, \quad Ab = \begin{bmatrix} 0 \\ \cdot \\ 0 \\ 0 \\ 1 \\ x \end{bmatrix}, \quad A^2 b = \begin{bmatrix} 0 \\ \cdot \\ 0 \\ 1 \\ x \\ x \end{bmatrix}, \quad A^{n-1}b = \begin{bmatrix} 1 \\ x \\ \cdot \\ \cdot \\ \cdot \\ x \end{bmatrix},$$

where x denoted elements with no concern. The representation (9′) is called an observable representation because n row vector $c', c'A', \dots, c'(A')^{n-1}$ have similar structure to allow ready determination of their linear independence. Controllable or observable state space representations are often used to simplify dynamic analysis.

Canonical forms. Matrices and vectors A, b and c introduced above are special because they have well-defined patterns of 0's and 1's and only a small number of non-zero elements that are related to model parameters. These state space representations are therefore known as canonical forms. We do not have a model in canonical state space form in practical applications. As we shall later discuss under (stochastic) realization theory, constructing canonical forms given time series data is sensitive to numerical errors. Noncanonical state space representation of data is generally more robust and may provide approximate representations of time series data by low-order state space models.

The process of putting time series models into state space form can be reversed, i.e. usual time series models can be recovered from dynamic systems in state space form. Start form y_t in (9)

$$y_t = c'z(t) + \beta_0 u_t ,$$

where the dimension of $z(t)$ is n. Advance times by one to relate y_{t+1} to $z(t)$ as

$$\begin{aligned} y_{t+1} &= c'x(t+1) + \beta_0 u_{t+1} \\ &= c'Az(t) + c'bu_t + \beta_0 u_{t+1} . \end{aligned}$$

Here the state vector transition equation (1st equation in (9)) has been used. Proceed analogously to generate successively

$$\begin{aligned} y_{t+2} &= c'A^2 z(t) + c'Abu_t + c'bu_{t+1} + \beta_0 u_{t+2} , \\ &\vdots \\ y_{t+n} &= c'A^n z(t) + c'An{-}1bu_t + \cdots + \beta_0 u_{t+n} . \end{aligned}$$

Now the characteristic polynomial of the matrix A introduced in (2) is

$$|\lambda I - A| = \lambda^n + \alpha_1 \lambda^{n-1} + \cdots + \alpha_n .$$

Multiplying $y_t, \ldots, y_{t+n-1}$ by $\alpha_n, \ldots, \alpha_1$ and adding them together, we can eliminate $z(t)$ from $y_t, \ldots, y_{t+n}$ because $c'(A^n + \alpha_1 A^{n-1} + \cdots + \alpha_n I)z(t) = 0$ by the Cayley Hamilton theorem, i.e. $y_{t+n} + \alpha_1 y_{t+n-1} + \cdots + \alpha_n y_t$ now involves only u's on the right, thus showing that sequences of y's and u's are related by an ARMA model.

The difference equation for $z(t)$ is stable if and only if distinct eigenvalues of A are not greater than one in modulus and repeated eigenvalues have modulus strictly less than one. Comparing the expression for the characteristic polynomial with $\emptyset(L)$, we note that $1/L$ corresponds to λ. The usual stability condition for $\emptyset(L)$ is then recovered from the conditions on the magnitudes of the eigenvalues by noting this correspondence.

So far, coefficients α's and β's are implicitly treated as constants. Our derivation, however, need not assume them to be constants. They can be varying with time. The only complication from time-varying α's and β's is that the matrices (and vectors) A, b and c become time-varying.

Time series models with time-varying parameters can thus be put in state space form just as easily as time-invariant dynamic systems.

Dynamic Multiplier Calculation

Earlier we claimed that state space representation leads to simpler statements of dynamic model properties. To illustrate this, let us calculate impulse responses or dynamic multipliers of a time series model using the state space representation. They are defined as (partial) derivatives of y_t with respect to u_τ, $\tau < t$. Dynamic multipliers are important in policy effectiveness studies because they tell us how policy effects on y vary with time.

In the state space form (9) the difference equation for the state vector can be solved as

$$z(t) = A^t z(0) + \sum_{\tau=1}^{t} W_\tau u_{t-\tau},$$

where

$$W_\tau = A^{\tau-1} b.$$

In this form it is almost self-evident that the multipliers are given by

$$\partial y_t / \partial y_{t-\tau} = c' W_\tau, \qquad 1 \leqslant \tau \leqslant t$$

and

$$\partial y_t / \partial u_t = \beta_0 .$$

This observation straightforwardly generalizes to a time series with vector-valued y's and u's.[4] Compare this straightforward derivation with a complicated and cumbersome time series approach typified by Brissimis [13].

Kalman Filter

Kalman filters provide another illustration of representational and conceptual simplicity gained by the state space approach. Suppose that a time series model is already in a state space form

$$x(t+1) = Ax(t) + \xi(t)$$

and

(10) $$y_t = Cx(t) + \eta(t) .$$

Here we have allowed for vector-valued $\{y_t\}$ processes.

Assume that $\xi(t)$ and $\eta(t)$ are jointly mean zero Gaussian processes, and that

$$E \begin{bmatrix} \xi(t) \\ \eta(t) \end{bmatrix} [\xi(s)', \eta(s)'] = \begin{bmatrix} Q_t & S_t \\ S_t' & R_t \end{bmatrix} \delta_{t,s} ,$$

where $\delta_{t,s}$ is one for $t = s$, and zero otherwise. The initial condition $x(0)$ is also assumed to be independent of all ξ's and η's.

The y-process is observed, but $x(t)$ is not directly observed. The problem is to forecast y_{t+T}, $T \geqslant 1$ based on the observation data $y_0, y_1, \ldots, y_t$. Denote by $y_{t\,|\,t-1}$ the conditional mean of y_t given $y_0, y_1, \ldots, y_{t-1}$. Define $v_t = y_t - y_{t\,|\,t-1}$. We write $\hat{x}(t)$ for $E(x(t)\,|\,y_0, \ldots, y_{t-1})$. Thus $v_t = C(x(t) - \hat{x}(t)) + \eta(t)$. The process $\{v_t\}$ is known as an innovation process because v_t is orthogonal to the subspace spanned by past observation data $y_0, \ldots, y_{t-1}$, i.e. independent of these vectors, and representing 'new' information contained in the most recent observation y_t.

Because of the independence assumption on disturbances, $y_{t+T\,|\,t}$ equals $A^{T-1}y_{t+1\,|\,t}$. The prediction problem thus reduces to calculating the one-step ahead prediction $y_{t+1\,|\,t}$ from (10).

From (10) we can show that

$$\hat{x}(t+1) = A\{\hat{x}(t) + K_t v_t\}$$

and

$$(11) \qquad y_{t+1\,|\,t} = C\hat{x}(t)$$

for appropriately chosen K_t. The Appendix establishes that

$$K_t = (S_t + P_t C')(R_t + CP_tC')^{-1},$$

where the estimation error covariance is

$$P_t = \operatorname{cov} \hat{x}(t),$$

and where we assume that R_t is nonsingular.

Denote the conditional estimation error covariance matrix by

$$\Gamma_t = \operatorname{cov}(x(t)\,|\,y_0, \ldots, y_t).$$

Then Γ_t is related to P_{t+1} by

$$P_{t+1} = Q_t + A\Gamma_t A'.$$

The recursion formula relating P_{t+1} to P_t is known as the Riccati equation.[5]

This sequential representation has computational as well as analytical advantages. The sequential updating scheme is computationally useful since new observations can be incorporated as they arise. To illustrate the conceptual or analytical advantages, consider the effects of P_0 on the filter gain calculation. (11) shows that K_t depends on P_0 among other things. Since P_0 is unlikely to be known in practice, it is of interest to examine conditions which ensure that K_t approach a constant K independent of P_0 as t becomes large. The condition for the existence of asymptotic time invariant filter gain can be succinctly put in term of the concepts of controllability and observability or their weaker version of stabilizability and detectability, or more directly as the condition of $A - \bar{K}C$ being asymptotically stable where $\bar{K} = \lim K_t$. The proof of this fact is straightforward by appealing to the 'duality' between the estimation problem and the regulation problem (see [21] or [9], Chapter 5, 9, 10). Such conceptual framework has not been available in the traditional time series literature.

Kalman filters, especially with the gain matrix which need not be optimal, can subsume a host of ad-hoc filtering or smoothing schemes which abound in the econometric literature. Alternatively, Kalman filters can be used to suggest asymptotically optimal filtering schemes. A popular exponential forecasting scheme illustrates this. Given sequence of observations $y_1, y_2, \ldots, y_t$ on $x_1, x_2, \ldots, x_t$, a forecast of x_{t+1} is formed by $\hat{x}_{t+1} = \alpha y_t + \alpha(1-\alpha)y_{t-1} + \alpha(1-\alpha)^2 y_{t-2} + \cdots$ for some $0 < \alpha < 1$. This is equivalent to $\hat{x}_{t+1} - (1-\alpha)\hat{x}_t = \alpha y_t$, i.e. $\hat{x}_{t+1} = (1-\alpha)\hat{x}_t + \alpha y_t = \hat{x}_t + \alpha(y_t - \hat{x}_t)$ is an example of the exponential smoothing scheme.

Suppose y_t and x_t are related by $y_t = x_t + \eta_t$ and x_t evolves with time according to $x_{t+1} = x_t + \zeta_t$. Then the Kalman filter for this problem is

$$\hat{x}_{t+1} = \hat{x}_t + k_t(y_t - \hat{x}_t),$$

where

$$k_t = (\sigma_t^2 + s_t)/(\sigma_t^2 + r_t^2)$$

and

$$\sigma_t^2 = \text{cov}(x_t^2),\, s_t = E(\zeta_t \eta_t) \quad \text{and} \quad r_t^2 = E(\eta_t^2).$$

Only when k_t approaches a constant α, can the expotential smoothing scheme be justified as asymptotically optimal. In other words the coefficient α should be chosen to be the limit of k_t as t becomes infinite.

Duality in Dynamic Systems

The notion of duality is well known in linear and nonlinear programming literature. That certain static constrained maximization problems may alternatively be posed as static constrained minimization problems has proved useful both conceptually and computationally. Realization that a certain class of dynamic optimization problems, known as quadratic cost-regulator problems in the systems literature, is 'dual' to the Kalman filtering problem constitutes a dynamic extension of this powerful notion of duality. Dynamic duality does not refer to alternative formulations of a given optimization problem as maximization or minimization problems, but rather the notion points out formal (mathematical) similarities in two distinct classes of optimization problems.

This relationship between the two seemingly unrelated classes of problems for dynamic systems provides a conceptual framework not available in the traditional time series literature and have proved quite helpful in establishing asymptotic properties for the estimation error covariance matrices, for example. The quadratic-cost regulator problem is to minimize a quadratic expression of the outputs and controls such as $\Sigma_{t=1}^{T} y_t' y_t$ subject to the dynamics in its state space representation such as $z_{t+1} = Fz_t + Gu_t$, $y_t = Dz_t + Eu_t$. The solution involves a Riccati equation just as the Kalman filter problem does. More precisely, the correspondence $F \leftrightarrow A'$, $G \leftrightarrow C'$ and others involving noise covariance statistics, D and E, cause the Riccati equations for calculating the filter gain and gain for the feedback control to be identical (see [9], Chapters 5 and 9 for details). The asymptotic limit of the one-step-ahead prediction error covariance can then be easily established by solving the dual regulator problem as T approaches infinity. The latter can be shown to exist if certain observability conditions are met for a fictitious dynamic system.

The concepts of observability or controllability thus turn up as important technical conditions in dynamic optimization problems, in addition to being important dynamic properties *per se.*[6]

Deterministic Realization

To convey the basic idea we assume that a sequence of impulse responses of a dynamic model admits a finite dimensional state space representation, i.e. finite dimensional matrices A, B and C exist such that the sequence $\{H_k\}$ can be represented by $H_k = CA^{k-1}B$, $k = 1, 2, \ldots$, i.e. the Hankel matrix can be factored as follows

$$H = \begin{bmatrix} H_1 & H_2 & H_3 & \cdots\cdots \\ H_2 & H_2 & \cdots\cdots & \\ H_3 & \cdots & \cdots\cdots & \\ \cdots & \cdots & \cdots\cdots & \end{bmatrix} = \begin{bmatrix} C \\ CA \\ \vdots \end{bmatrix} [B, AB, \ldots]$$

Denote the first factor by $\mathcal{O}$ (observability) and the second by $\mathcal{C}$ (for controllability). Theoretically the rank of H is n which is the dimension of the controllable and observable state space realization.

Shift the semi-infinite matrix $\mathscr{C}$ to the left by r, the dimension of the input vector and denote it by $\mathscr{C}_1$ or

$$\begin{aligned}\mathscr{C}_1 &= A[B, AB, \ldots] \\ &= A\mathscr{C}.\end{aligned}$$

So numerically A is recovered as $A = \mathscr{C}_1 \mathscr{C}^+$ where $\mathscr{C}^+$ denotes the pseudo-inverse. The left r columns give B and the top m rows of $\mathscr{O}$ define C where m is the dimension of the output vector. The matrix A can similarly be recovered by up-shifting $\mathscr{O}$. By singular decompositions of H as $U\Sigma V$ and identifying $\mathscr{O}$ with $U\Sigma^{1/2}$ and $\mathscr{C}$ by $\Sigma^{1/2}V$, we see that $A = \Sigma^{1/2}U^*U\Sigma^{1/2} = \Sigma^{1/2}VV^*\Sigma^{-1/2}$. This produces a realization known as internally balanced, which has some desirable properties [9a] [23]. Hankel matrices also admit approximate factorization by retaining only the largest k singular values out of possible n if k is specified [19].

Stochastic Realization

How do we put time series data in state space form to take advantage of alternative ways of predicting future values of time series by Kalman filters? This problem is known as stochastic realization (see [2] or [14] for some references). In its basic form the problem is to find a Markovian representation for a given sequence of stationary covariance matrices. Deterministic realization theory produces a state space representation from a sequence of input-output data. Stochastic realization produces noise covariance matrices given dynamics in state space form. Consider

$$\begin{aligned}x_{t+1} &= Fx_t + v_t \\ y_t &= Hx_t + w_t,\end{aligned}$$

where u_t and w_t are stationary mean 0 independent stochastic processes, with

$$\operatorname{cov}\begin{bmatrix} v_t \\ w_t \end{bmatrix} = \begin{bmatrix} Q & S \\ S' & R \end{bmatrix} \geqslant 0 .$$

Let $P = E(x_t x_t')$. Then from stationarity, P satisfies

$$\begin{aligned} P &= E(x_{t+1} x_{t+1}') \\ &= FPF' + Q . \end{aligned}$$

We note that the covariance matrix of the output y_t is

$$\Lambda_0 = E(y_t y_t') = HPH' + R$$

and

$$\Lambda_T = E(y_{t+T} y_t') = HF^{t-1} G ,$$

where $G = FPH' + S$.

The problem is to find $P \geqslant 0$, Q, S, and R such that

$$\begin{bmatrix} Q, & S \\ S' & R \end{bmatrix} \geqslant 0 ,$$

satisfying the above equations given a sequence $\{\Lambda_T\}$. This problem amply illustrates the power of the system theoretic approach because duality of the regulator and filtering problem can be used to advantage in establishing the existence of the solution to the stochastic realization problems [14].

STATE SPACE MODELLING OF INTERDEPENDENT ECONOMIES

In this section we shall illustrate how models in state space form help sort out complicated dynamic effects of policy instruments on interdependent national economies by combing variational dynamic analysis, state space representation, and Laplace transforms.

Comparison of alternative policy régimes can be made by dynamically evaluating effects of policy differences on endogenous variables, i.e. by calculating incremental dynamic policy multipliers of appropriate pairs of endogenous variables and instruments.

More specifically we show that (incremental) dynamic policy multipliers can be calculated almost as simply as calculations of impact or long-run policy multipliers, if Laplace transforms are used.

The ability to sort out complicated interdependent policy effects is the attribute emphasized in this section, rather than 'mere' predictive ability or realism of models. Material is drawn from Aoki [3, 4, 6], to which the reader is referred for a more complete and leisurely exposition.

Variational Model is a Detrended Log-Linear Model

Consider a model in an abstract structural form

$$Z(t) = F(Z(t), X(t), t; \theta),$$

where Z is a state vector, X is a set of exogenous (policy) variables and θ is a vector of model parameters.

A basis for comparison, that is the 'control' or reference solution, is denoted by superscript 0,

$$Z^0(t) = F(Z^0(t), X^0(t); t; \theta^0).$$

Generally, $Z^0(t)$ is changing with time unless a long-run equilibrium state is chosen as a basis for comparison.

For each component of the vector we can introduce relative deviation defined by

$$z_j(t) \cong \{Z_j(t) - Z_j^0(t)\}/Z_j^0(t),$$

unless Z_j^0 is zero, in which case $Z_j(t) - Z_j^0(t)$ is denoted as z_j.

Taking the logarithm of $Z_j = Z_j^0(1 + z_j)$ on the assumption that it is positive, and assuming further that $|z_i|$ is rather small, we note that $z_j \simeq \ln Z_j - \ln Z_j^0$. When $Z_j^0(t)$ is growing exponentially such as when a balanced growth path is chosen as a reference path, then z_j denote the usual detrended variable, often encountered in the economics literature. The collection of $z_j(t)$ form a state vector governed by the differential equation

$$(12) \quad \dot{z}(t) = A(t;\theta)z(t) + B(t;\theta)x(t).$$

This equation is called the variational dynamic equation (about the reference path). It is the same as a detrended log-linear model often used in econometric works. A special but a useful case is when a balanced growth path is chosen as the reference because the time-varying matrices A and B then become constant if intensive variables are used as original Z's (see Chapters 1 & 2 of [6], and [8]). The matrices $A(t;\theta)$ and $B(t;\theta)$ also often simplify for some special values of θ, say θ^*.

The solutions of (12) can be approximately constructed as $z^*(t) + \delta z(t)$ from the solution of the differential equation

$$\dot{z}^*(t) = A(t;\theta^*)z^*(t) + B(t;\theta^*)x(t)$$

and of the associated 'perturbation' equation which governs the time evolution of the difference of $z(t)$ and $z^*(t)$,

$$\delta\dot{z}(t) \simeq A(t;\theta^*)\delta z(t) + \delta A(t)z^*(t) + \delta B(t)x(t),$$

where

$$\delta z(t) = z(t) - z^*(t), \qquad \delta A = A(t;\theta) - A(t;\theta^*),$$

and

$$\delta B(t) = B(t;\theta) - B(t;\theta^*).$$

The usefulness of this procedure of course rests on the 'fact' that the dynamics with θ^* stands for a special case which is more easily analysed than the model with general θ. In a multiple-country world models, judicious choice of θ^* and the reference path often makes possible analytical treatment of complex issues. See [3], [7] and [8] for some examples.

Incremental Policy Multiplier

The solution of (12) is expressible as

$$z(t) = \phi(t, t_0)z(t_0) + \int \phi(t, \tau)B(t)x(t)\mathrm{d}\tau$$

where

$$\frac{\mathrm{d}}{\mathrm{d}t}\phi(t, t_0) = A(t; \theta)\phi(t, t_0), \ \phi(t_0, t_0) = I$$

is the fundamental solution matrix.

The first term evaluates the effects of displacement of initial conditions, and the second calculates the effects of changes in instruments [7].

This second term represents the dynamic effects of all past policy changes on the state vector. Variational or incremental changes in the time paths of target vector are related to $z(t)$ by

$$(13) \qquad q(t) = C(t; \theta)z(t) + D(t; \theta)x(t).$$

The impact effects are captured by the matrix $D(t; \theta)$. Its $(i - j)$th element is the impact multiplier in expressing the simultaneous affect of the change $x_j(t)$ on $q_i(t)$, i.e. $\{\Delta Q_i(t)/Q_i^0(t)\}/\{\Delta X_j(t)/X_j^0(t)\}$. The dynamic effects are given by

$$C(t; \theta)\int_{t_0}^{t} \phi(t, \tau)B(\tau)x(\tau)\mathrm{d}\tau.$$

Now, if the matrices A, B, C and D are all constant, then Laplace transforms can be used to our advantage.

Denote Laplace transforms by '~', $\tilde{z}$ for example. Form (12), ignoring the initial conditions, the Laplace transforms of the state vector and of the instrument vector is related by

$$\tilde{z} = (sI - A)^{-1} B\tilde{x} ,$$

and form (13), the Laplace transform of the target vector, is related to that of the instrument vector by

$$\tilde{q} = \{D + c(sI - A)^{-1} B\}\tilde{x} .$$

The first term is the matrix of impact multipliers while the second is the matrix of dynamic multipliers in Laplace transform. Note that the second term is calculated just as easily as the first, once the effects of incremental changes in the time paths of exogenous variables are specified as a (variational) state tradition equation and as a (variational) target Equation (13).

APPENDIX

Calculation of Kalman filer gain

Let

$$e(t) = x(t) - \hat{x}(t) .$$

From (10) and (11)

$$\begin{aligned} e(t+1) &= Ae(t) + \zeta(t) - AK_tCe_t - AK_t\eta(t) \\ &= A(I - K_tC)e(t) + \zeta(t) - AK_t\eta(t) . \end{aligned}$$

Let $\operatorname{cov} e(t) = P_t$. Noting that $e(t)$, $\zeta(t)$ and $\eta(t)$ are independent, the error covariance propagates according to

$$\begin{aligned} P_{t+1} &= A(I - K_tC)P_t(I - KC)'A' + Q_t \\ &\quad + AK_tR_tK_t'A' - S_tK_t'A' - AK_tS'_t . \end{aligned}$$

The optimal choice of K_t which minimizes the P_{t+1} must satisfy

$$A\{-(I-K_tC)P_tC'+K_tR_t-S_t\}\Delta K_t'A'=0$$

for arbitrary ΔK_t, or

$$K_t=(S_t+P_tC')(R_t+CP_tC')^{-1}.$$

Substitute this back into P_{t+1} to obtain the recursion form P_t to P_{t+1}.

NOTES

1 See Aoki [6, 7, 9] for detailed exposition of properties of models in state space representation and examples of dynamic models in state space form. Harvey [16] has a chapter on state space use in time series models.

2 Swamy and Tinsley [24] state that state representation of models facilitate derivation of identifiability conditions and parameter estimators, for example.

3 The method straightforwardly generalizes to vector-valued y's and u's. See Appendix A of [6] for example.

4 See note 3.

5 For example, if S_t is zero, then

$$P_{t+1}=Q_t+A[P_t-P_tC'(R_t+CP_tC')^{-1}CP_t]A',$$

where $(X^{-1}+C'Y^{-1}C)^{-1}=X-XC'(Y+CXC')^{-1}CX$ is used. This identity is true wherever the indicated inverse exists.

6 See [9], p. 114 for an example in which observability implies the unique market equilibrium.

7 We can consider both small changes in magnitude, i.e. $|x(\tau)|$ small for all $t_0\leqslant\tau$, or changes over a small time interval, i.e. with a small positive ϵ and arbitrary $\bar{t}$ and ζ,

$$x(t)=\begin{cases}\zeta, & t_0\leqslant\bar{t}-\epsilon\leqslant t\leqslant\bar{t}\\ x^0(t) & \text{otherwise.}\end{cases}$$

These two types of variation are known as weak and strong respectively. See [15] for examples.

REFERENCES

[1] A. B. Abel and O. J. Blanchard, 'An Intertemporal Equilibrium Model of Savings and Investments', Discussion Paper No. 739, Harvard Institute of Economic Research (1980.

[2] H. Akaike, 'Canonical Correlation Analysis of Time Series and the Use of an Information Criterion', in R. K. Mehra and D. G. Lainiotis (eds.), *System Identification: Advances and Case Studies* (Academic Press, New York, 1976).

[3] M. Aoki, 'Dynamic Distributive Effects of Money Stock Changes in a Three Country World Model under Flexible Exchange Rate Regimes', Discussion Paper No. 113, Institute of Social and Economic Research, Osaka University (1982).

[4] M. Aoki, 'On Existence of Exchange Rate Union in a Three-Country World under Flexible Exchange Rate Regimes', Discussion Paper No. 117, Institute of Social and Economic Research, Osaka University (1982).

[5] M. Aoki, 'A Basic Inequality for Dynamic Effects of Monetary and Real Distrubances: Dynamic Criterion for an Exchange Rate Union in a Three-Country World', Discussion Paper No. 122, Institute of Social and Economic Research, Osaka University (1982).

[6] M. Aoki, *Dynamic Analysis of Open Economies* (Academic Press, New York, 1981).

[7] M. Aoki, 'Note on Comparative Dynamic Analysis', *Econometrica* **48** (1980), 1319–25.

[8] M. Aoki, 'Comparative Dynamic Analysis of a Growth Model under Alternate Policy Regimes', *Journal of Macroeconomics* **2** (1980), 1–40.

[9] M. Aoki, *Optimal Control and System Theory in Dynamic Economic Analysis* (North-Holland American Elsevier, New York, 1976).

[9a] M. Aoki, *Notes on Economic Time Series Analysis: System Theoretic Perspectives* (Springer-Verlag, Berlin and New York, 1983).

[10] O. J. Blanchard, 'Debt and the Current Account Deficit in Brazil' Discussion Paper No. 865, Harvard Institute of Economic Research (1981).

[11] M. Blaug, *The Methodology of Economics* (Cambridge University Press, London and New York, 1980).

[12] G. E. P. Box and G. M. Jenkins, *Time Series Analysis Forecasting and Control* (Holden-Day, San Francisco, 1976).

[13] S. Brissimis, 'Multiplier Effects for Higher than First-Order Linear Dynamic Econometric Models', *Econometrica* **44** (1974), 593–5.

[14] P. Faurre, 'Stochastic Realization Theory', in R. K. Mehra and D. G. Lainiotis (eds.). See [2].

[15] W. Fleming and R. W. Rishel, *Deterministic and Stochastic Optimal Control* (Springer-Verlag, New York, 1974).
[16] A. C. Harvey, *Time Series Models* (Phillip Allan, Oxford, UK, 1981).
[17] L. Johnasen, 'Econometric Models and Economic Planning and Policy: Some Trends and Problems', Mimeo (Institute of Economics, University of Oslo, 1981).
[18] M. L. Kac, 'Some Mathematical Models in Science', *Nature* **166** (1969), No. 3806, 695–9.
[19] S-Y. Kung and D. W. Lin, 'Recent Progress in Linear System Model-Reduction Vic Hankel Matrix Approximation', Mimeo (Department EE, USC).
[20] B. C. Moore, 'Principal Component Analysis in Linear System: Controll-ability, Observability and Model Reduction', *IEEE Transactions Automatic Control* **AC–26** (1981), 17–32.
[21] E. Mosca and G. Zappa, 'Consistency Conditions for the Asymptotic Innovation Representation and an Equivalent Inverse Regulation Problem', *IEEE Transactions Automatic Control* **AC–24** (1979), 63–9.
[22] M. Obstfeld, 'Macroeconomic Policy and World Welfare under Flexible Exchange Rates', Mimeo (Department of Economics, Columbia University, 1980).
[23] L. Pernebo and L. M. Silverman, 'Balanced Systems and Model Reduction', Proceedings of the 18th IEEE Conference on Decision and Control, (1980), pp. 865–7.
[24] A. V. B. Swamy and P. A. Tinsley, 'Linear Prediction and Estimation Methods for Regression Models with Stationary Coefficients', *Journal of Econometrics* **12** (1980), 103–42.

Institute of Social and Economic Research, Osaka University, Japan

J. C. Willems

Systems Sciences and Physics

DESCRIPTIVE VERSUS PRESCRIPTIVE SCIENCES

The popular image of a scientist is heavily biased towards the *descriptive sciences* (physics, chemistry, astronomy, etc.). One pictures a scientist as an observer – preferably one who is doing experiments – summarizing our understanding of what is happening around us in Nature in the form of laws which will hopefully help us preview what will be happening if similar future situations should occur or be induced to occur. The central role of *falsification* in the philosophy of science strongly supports such an image. Also, when a scientific activity like mathematics does not quite fit within this framework, its existence is likely to be justified on the basis that it has its historical roots in the need to provide a suitably precise language to support the descriptive sciences. This historical conception of science should not surprise us. Indeed, physics – the descriptive science *par excellence* – has permanently dominated the scientific scene. It is even today the most influential science and has undoubtedly attracted the finest scientific minds.

The need for *prescriptive sciences* was first felt as a consequence of the industrial revolution. Out of this, engineering emerged as a new discipline. Now, the questions which one asks in engineering are in principle questions of design. This involves coming up with a procedure for achieving a particular goal, for example the problem of connecting a number of components into a system (say a mechanical machine or an electrical circuit) which performs a particular function. Of course, in design one has to take into consideration the constraints imposed by the building blocks used. These could be mechanical parts, electrical devices, chemical reactions, etc. For the understanding of these constraints engineering was able to draw on extensive knowledge in the descriptive sciences. However, the design process also involves coming up with a *blueprint* for interconnecting the building blocks. This is often referred to as *synthesis*, in contrast to *analysis* by which we mean the descriptive aspect, i.e. understanding the constraints imposed by the laws of the building blocks. However, in conventional engineering this synthesis process never really achieved a very strong scientific basis, and it has always remained more of an art than a science.

This situation has changed considerably in the last two decades with the advent of computer science and the many engineering tasks which involve the computer in one way or another. Indeed, this has radically shifted the relative importance of understanding the constraints versus the blueprint, i.e. the '*synthesis*' versus the '*analysis*'. In computer science, the constraints imposed by the physical reality of the electronics of the calculating machine take a clear second place compared with the complexity of planning, in the form of a logical computer code, the task to be performed.

An example of a scientific endeavour which has crystallized out of the prescriptive aspect of engineering is what is called (*mathematical*) *system theory*. This field has its roots in questions of control engineering and analogue data processing, as filtering for example. Now the task of designing a control system or a filter quite clearly involves synthesis. It is interesting to observe that it took quite some time before control questions were actually phrased as synthesis problems. Indeed, in the early contributions by Maxwell, Lyapunov, etc. the emphasis was entirely on analysis. In the contributions of the late 1930s by Nyquist, Bode *et al.* there was already an implicit element of synthesis. However, it was really only about twenty-five years ago that the synthesis question came to dominate the control scene, through clearly formalized synthesis questions in optimal control, least squares filtering, etc.

From what we have said before it is clear that the philosophy and the roots of the problems in *physics*, the queen of the descriptive sciences, and *systems theory*, one of the more developed areas in the descriptive sciences, are fundamentally different. Notwithstanding this, both fields show a great deal of affinity and similarity. The reason is that both consider rather similar dynamical objects: they are both heavily involved with dynamical systems described by (ordinary or partial) differential equations. In fact, it is precisely this similarity combined with the great difference in philosophical vantage point which makes a successful cross-fertilization possible. We hope to illustrate this in the remainder of this article.

THE CONCEPT OF A DYNAMICAL SYSTEM IN PHYSICS

In physics one looks at a dynamical system from the *inside*. By this we mean that the demands on a good model require that it should be an explanatory one: not only should it give a good description of the behaviour of a system but it is also expected to provide an explanation of this behaviour in terms of some simple, generally accepted principles. However, systems are usually influenced by their environment and in most cases it will be impossible to model or predict this influence. In order to circumvent this difficulty, physics (or at least the mathematical axiomatication of physics) seems almost invariably to have gone for the assumption that the outside world remains constant in some sense.

Let us illustrate this in the context of *Hamiltonian mechanics*, which is believed to capture what we intuitively think of as mechanical systems. In this theory one postulates that such a system can be described by its state x and that the collection of all possible states can be organized as a symplectic manifold. In addition, the evolution of x is assumed to be governed by a Hamiltonian vector field. More concretely, this means that one can choose (locally) canonical coordinates (p, q) for x, with $p \in \mathbb{R}^n$, the *momentum*, and $q \in \mathbb{R}^n$, the *configuration*, and define a function $(p, q) \mapsto H(p, q)$, with H the *Hamiltonian* (its physical significance is that of total energy). In terms of this, the evolution of (p, q) is then governed by the differential equations

$$\dot{q} = \frac{\partial H}{\partial p}(p, q), \qquad \dot{p} = -\frac{\partial H}{\partial q}(p, q).$$

This same structure can be found in *quantum mechanics*, where one postulates that the state of a particle can be described by its wave function $\psi : \mathbb{R}^3 \to \mathbb{C}$ such that $\psi \in L_2$. The time evolution of ψ_t is then governed by

$$i \frac{h}{2\pi} \frac{\mathrm{d}\psi_t}{\mathrm{d}t} = H\psi_t,$$

with h Planck's constant and H a self-adjoint operator on L_2.

Under suitable conditions both the above situations define a uniquely solvable initial value problem which induces a transformation from the initial state at time 0 to the state at time t, i.e. $(p, q)(0) \mapsto (p, q)(t)$ and $\psi_0 \mapsto \psi_t$. All this may be formalized a great deal further. However, these examples already allow us to make our point. In the above examples it is assumed that the state transformation is a completely deterministic one. i.e.

the possibility that the conditions at the boundary of the system may vary is not explicitly taken into account. It is assumed that these boundary conditions remain constant or at least (if one assumes that H also depends on time explicitely) that they are known functions of time.

THE CONCEPT OF A DYNAMICAL SYSTEM IN SYSTEMS THEORY

In systems theory on the other hand one looks at a dynamical system from the *outside* of the system. From the point of view of prescriptive sciences, this vantage point is a must since the main interest is how the system can be used, i.e. how it will react to outside stimuli. This implies that one may be content with a 'black box' model: a model which predicts the behaviour of a system but which has no pretences about explaining it. A typical class of models of this sort is one where the input $u(\cdot)$ of a system is related to its output $y(\cdot)$ by means of the convolution integral $y(t) = \int_{-\infty}^{t} g(t - \tau)u(\tau)\,d\tau$. The kernel of this convolution, the function $g(\cdot)$, describes the dynamics of the system. Thus one views a system as a law which transforms the external input which is imposed on it from the outside into the external output which then shows how the system in its turn influences its environment.

An important notion both in physics and in systems theory is that of the *state* of a system. Intuitively, the present state of a system should contain sufficient information such that it determines its future behaviour. We have already seen that in Hamiltonian mechanics the state consists of the canonical variables (p, q) while in quantum mechanics it consists of the wave function ψ. In these situations the initial state determines the entire state trajectory uniquely. However, in systems theory it cannot be the case that the state yields uniquely the future trajectory because this trajectory will also depend on the (unknown) input. Nevertheless, the notion of state may be generalized to the input/output setting. The idea is to choose the state variables such that they contain sufficient information about the past so that the map which determines the future output on the basis of the future input will depend only on the present state (and not on the rest of the past input). A typical state space model then takes the form of a set of first-order differential equations: $\dot{x} = f(x, u), y = g(x)$, with u the input, y the output, and x the state.

Before discussing why the framework of systems theory also provides a useful context for the description of physical systems, let us formalize, and to some extent generalize, all this. The first point to note is that in physical systems it is often not clear *a priori* what are the input and what are the output variables (in fact, in some situations it may not even be possible to divide the external variables into input and output variables). It is therefore much more useful to start from a definition of a dynamical system which models the external behaviour without making a distinction between inputs and outputs. These considerations have led to the following definition [1]. With a (continuous time) *dynamical system* we mean a triple $\Sigma_e = \{\mathbb{R}, W, \mathcal{S}_e\}$ with W the *signal alphabet* (the space where the signals which we are trying to model take on their values) and $\mathcal{S}_e$ is a family of maps from the time axis $\mathbb{R}$ to W. $\mathcal{S}_e$ is called the (*external*) *behaviour* of Σ_e. Thus a signal $w(\cdot) : \mathbb{R} \to W$ belongs to $\mathcal{S}_e$ iff it is compatible with the laws governing the system. With a *state space system* we mean a quadruple $\Sigma_i = \{\mathbb{R}, W, X, \mathcal{S}_i\}$, with X the *state space*, and $\mathcal{S}_i$ a family of maps from $\mathbb{R}$ into $X \times W$, called the *internal behaviour* of Σ_i. It is assumed that $\mathcal{S}_i$ satisfies the *axiom of state*, that is to say that if (x_1, w_1) and (x_2, w_2) are two possible trajectories in $\mathcal{S}_i$ with $x_1(t) = x_2(t)$, then their t-concatenation, $(x_1, w_1)^- \wedge_t (x_2, w_2)^+$ will also belong to $\mathcal{S}_i$. This t-concatenation is defined by

$$((x_1, w_1)^- \underset{t}{\wedge} (x_2, w_1)^+)(\tau) := \begin{cases} (x_1, w_1)(\tau) & \text{for} \quad \tau < t \\ (x_2, w_2)(\tau) & \text{for} \quad \tau \geqslant t. \end{cases}$$

The class of systems where the initial state determines the whole future trajectory (as in the situations discussed in the previous section) are called autonomous systems. Thus a state space system is called *autonomous* if there exists a map f: $x(t) \mapsto (x(\cdot), w(\cdot))$ such that $\{(x, w) \in \mathcal{S}_i\} \Longleftrightarrow \{(x, w) = f(x(t))\}$.

It is easy to define the *external behaviour* of a given state space system $\Sigma_i = \{\mathbb{R}, W, X, \mathcal{S}_i\}$. This yields $\Sigma_e = \{\mathbb{R}, W, \mathcal{S}_e\}$ with $\mathcal{S}_e := \{w \mid \exists x$ such that $(x, w) \in \mathcal{S}_i\}$. We say that Σ_e is *induced* by Σ_i. The converse problem, i.e. to derive for a given system Σ_e a state space system Σ_i such that Σ_i induces Σ_e is called the problem of *realization* (or the *state space representation problem*). The state space representation problem requires, among other things, coming up with a suitable state space X, which makes the problem a non-trivial one, particularly if one demands it to have some nice additional properties or structure.

THE ROLE OF EXTERNAL VARIABLES IN PHYSICS

The situation to which Hamiltonian mechanics addresses itself, and where one assumes that the system is an autonomous one, may be suitable for celestial mechanics. However, for the mechanical systems encountered on earth autonomous systems provide an extremely limited context. Indeed, most mechanical constructs around us (such as bicycles, windmills, cars, etc.) would become inert and rather uninteresting objects were it not for the possibility of exerting external forces on them. Note furthermore that the assumption that this external force comes from a potential field will usually be an absurd one. Also, the point of view that the external force is a given time function (which may be taken into account by assuming the Hamiltonian to be time-varying) is not adequate either, since the object of primary interest is the *map* from these forces to the mechanical variables (e.g. the configuration) and one should not suppress the nature of this map by looking at its image at one point.

Similar objections may be put forward against other areas of physics (as quantum mechanics) where the axiomatization has led to autonomous systems. A reasonably complete model of a physical phenomenon should incorporate the external influences as basic variables.

These considerations lead to an important area of inquiry where the fields of systems theory and physics meet. The goal of this reasearch is to use the axiomatic view of dynamics as provided by systems theory as a starting point for studying dynamics in physics, and to incorporate into it the special structures which one encounters in physics.

This programme has already been carried out to some extent for Hamiltonian systems with external forces and observations [2, 3]. It leads to the following definitions. With a *differential dynamical system* we mean a state space system $\Sigma_i = \{\mathbb{R}, W, X, \mathcal{S}_i\}$, with W and X smooth manifolds and $\mathcal{S}_i$ defined by a smooth fibre bundle $\pi : B \to X$ and a map $f : B \to TX \times W$ such that $\pi = \pi_X \cdot f$. Here TX denotes the tangent bundle of X and π_X the canonical projection of $TX \times W$ on X. One then defines $\mathcal{S}_i$ as $\mathcal{S}_i := \{(x, w) : \mathbb{R} \to X \times W \,|\, x$ is absolutely continuous and $(\dot{x}(t), w(t)) \in f(\pi^{-1}(x(t)))$ for almost all $t\}$. We will now add structure to this, leading to a *Hamiltonian system*. This is done by asking that W and X are symplectic manifolds and that $f(B)$ is a Lagrangian submanifold of $TX \times W$ (considered as a symplectic manifold in the natural way). In terms of (local) coordinates this is equivalent to asking that we can choose canonical coordinates (p, q) in X and $((u_1, y_2), (y_1, u_2))$ in W such that the equations of motion can be described by means of a single function H :

$(p, q, u) \mapsto H(p, q, u)$, the *Hamiltonian*, such that the evolution of p and q in terms of u, and the value of y as function of p, q, and u, is given by:

$$\dot{q} = \frac{\partial H}{\partial p}(p, q, u_1, u_2), \quad \dot{p} = -\frac{\partial H}{\partial q}(p, q, u_1, u_2),$$
$$y_1 = \frac{\partial H}{\partial u_1}(p, q, u_1, u_2), \quad y_2 = -\frac{\partial H}{\partial u_2}(p, q, u_1, u_2).$$

In this representation are should interpret the (local) inputs u_1 and the (local) outputs y_2 as external forces and the (local) inputs u_2 and the (local) outputs y_2 as configuration variables. Thus the canonical states are positions and momenta, while the canonical inputs and outputs are force and configuration, on vice versa.

This elegant structure and representation has been used to obtain generalizations of the Lagrangian formulation of mechanics [3], a generalization of Noether's theorem on symmetries and conservation laws [4], etc.

THE ROLE OF THE REALIZATION PROBLEM IN PHYSICS

In physics one always basically assumes that it can be postulated from the nature of the problem what the state variables are (the configuration and the momenta in classical mechanics, the wave function in quantum mechanics, electrical and magnetic fields in electromagnetic theory, etc.). In a sense this can be considered as a major element in the postulates of classical mechanics, quantum mechanics, Maxwell's equations, etc. The modelling process itself then involves writing the differential equations which determine the time evolution of these state variables. Because of the nature of the state one has to make sure that the initial state determines the whole state trajectory. This implies that this evolution will be described by a set of first-order differential equations in the state.

In systems theory on the other hand one looks at this modelling procedure in quite a different way. Indeed, one assumes that certain variables are to be modelled and that the modelling leads to a specification of the external behaviour of a system. Since there are no a-priori postulates on these external variables, no special restrictions enter the modelling problem at this stage. However, the value of the external variables at some time will in most circumstances not contain enough information to specify the state of the system. If we want to write state equations, then we will need to introduce additional variables. The formalization of this procedure is what we have called the state space realization problem. The model which one ends up with is now a state space model, but the state space was *induced* on the basis of the external behaviour of the system.

This system theoretic procedure for arriving at a state space model is a much more appealing one from many points of view. For one thing, it proceeds in a more logical way, starting from any type of model based, say, on external observations. Secondly, it is not hampered by the fact that the state may depend in a very sensitive way on the accuracy with which we decide to model. Indeed, working with a slightly more accurate model may have a dramatic effect on the nature of the state space as an object itself. To see this, note for example that in celestial mechanics the assumption that the planets are very hard elastic spheres instead of point masses will completely change the nature of the state space. Indeed, from a finite dimensional space it will become an infinite dimensional one. These difficulties are not present (or are at least conceptually more easy to manage) if one starts with the external behaviour of a system. True, modelling will more accurately introduce more terms in the equations but these terms will be small. The fact that the state space may have become much larger will automatically become evident in the state space realization stage which is often a very discontinuous process.

Hence, starting with the *external behaviour* of a system is a very pleasing phenomenological starting point. It also induces a number of questions which have important applications in physics. These issues are the following: assume that the external behaviour of a system possesses a certain qualitative property, what is the corresponding qualitative property for the external behaviour? Is it true that the external behaviour inherits this qualitative property from the internal behaviour? Finally, is it true that, if the external behaviour of a system possesses a certain qualitative property, then there will exist a state space realization which also has this qualitative property? In examining these questions, it is useful to note that we always have a choice of coordinates' on the state space X. This results from the fact that if the state space system $\Sigma_i = \{\mathbb{R}, W, X, \mathscr{S}_i\}$ has external behaviour $\mathscr{S}_e$, then, for any bijection $S: X \to X$, $\Sigma_i' = \{\mathbb{R}, W, X, \mathscr{S}_i'\}$, with $\mathscr{S}_i' := \{(x', w) \mid x'(\cdot) = Sx(\cdot) \text{ and } (x, w) \in \mathscr{S}_i\}$, will also be a state space system with the same external behaviour. This choice of the coordinates in the state space and the ensuing non-uniqueness of the state is essential in understanding the above representation questions.

We will now illustrate these questions by two examples.

Time Reversibility [5]

Let $\Sigma_e = \{\mathbb{R}, W, \mathscr{S}_e\}$ be a dynamical system. Let S_Δ denote the Δ-*shift*, defined by $(S_\Delta f)(t) := f(t - \Delta)$ and R *time-reversal*, defined by $(Rf)(t) := f(-t)$. A dynamical system with shift invariant behaviour ($S_\Delta \mathscr{S}_e = \mathscr{S}_e$ for all $\Delta \in \mathbb{R}$) is said to be *time-reversible* if $R\mathscr{S}_e = \mathscr{S}_e$. This definition obviously extends to a state space system $\Sigma_i = \{\mathbb{R}, W, X, \mathscr{S}_i\}$ and we immediately see that the external behaviour of an internally time-reversible system is also time-reversible. However, the converse is not always true, i.e. a system which is externally time-reversible may not have internally time-reversible behaviour. Actually this fact should not surprise us: if we consider the external behaviour of the positions of a classical mechanical system, then its state will consist of the position and the momentum of the pointmasses. Now, in order to backtrack on a trajectory while reversing time, we will at least have to change the signs of the momenta. However, we can work with a slightly more general notion of time-reversibility. Let us define a system to be *quasi-time-reversible* when there exists an involution S on W (an involution S is a bijection with $S^2 = I$) such that $R\mathscr{S}_e = S\mathscr{S}_e$. Note that if W is a vector space and S is linear, then for a suitable choice of the basis an involution simply changes the sign of some of the coordinates. With this definition it is then possible to prove that a system is quasi-time-reversible iff it has a quasi-time-reversible realization.

Thermodynamics [6]

Let us define a system $\Sigma_e = \{\mathbb{R}, W, \mathscr{S}_e\}$ with shift invariant behaviour to be *dissipative* if for all periodic motions $W \in \mathscr{S}_e$ we have $\oint \rho(w(t))$ at $\geqslant 0$. Here $\rho: W \to \mathbb{R}$ denotes the *supply rate* (e.g. the instantaneous work done on the system). If $\oint \rho(w(t))\,\mathrm{d}t = 0$ then we will say that the system is *conservative*. We define a smooth state space system $\Sigma_i = \{\mathbb{R}, X, W, \mathscr{S}_i\}$ with shift invariant internal behaviour $\mathscr{S}_i$ to be *dissipative* with *storage function* $S: X \to \mathbb{R}$ and *supply rate* $\rho: W \to \mathbb{R}$, if for all $(x, w) \in \mathscr{S}_i$ there holds

$$\frac{\mathrm{d}S(x(t))}{\mathrm{d}t} \leqslant \rho(w(t)).$$

If equality holds then we say that the system is *conservative*. For this class of systems we can prove that a system $\{\Sigma_e, \rho\}$ is externally dissipative iff there exists a state space realization Σ_i and a storage function $S: X \to \mathbb{R}$ such that $\{\Sigma_i, S, \rho\}$ is internally dissipative.

These results allow us to express the first and second laws of thermodynamics in a logically consistent way. Consider, for example, a body at uniform temperature T which is heated at a rate Q and performs work at a rate W. The physical laws specify a family of feasible time functions $(T(\cdot), Q(\cdot), W(\cdot))$ which together define a shift invariant external behaviour $\mathcal{S}_e \subseteq (\mathbb{R}^+ \times \mathbb{R} \times \mathbb{R})^{\mathbb{R}}$. This yields an externally defined dynamical system $\Sigma_e = \{\mathbb{R}, \mathbb{R}^+ \times \mathbb{R} \times \mathbb{R}, \mathcal{S}_e\}$. Now, the first and second laws of thermodynamics constrain this external behaviour $\mathcal{S}_e$ by stating that with $\rho_1 = Q - W$ and $\rho_2 = -Q/T$, $\{\Sigma_e, \rho_1\}$ is conservative and $\{\Sigma_e, \rho_2\}$ is dissipative. Note that these statements do not involve states, entropy, or energy. In order to introduce these, consider a state space realization Σ_i of Σ_e with state space X. Then from what we have seen before we can conclude that there exist functions $E: X \to \mathbb{R}$, the *internal energy*, and $S: X \to \mathbb{R}$, the *entropy*, such that $\{\Sigma_i, E, \rho_1\}$ is conservative and $\{\Sigma_i, -S, \rho_2\}$ is dissipative. These statements then yield the more conventional forms of the first and the second law. However, the manner in which these laws are stated in thermodynamics are usually highly ambiguous. Indeed, in the absence of a clear axiomatic state space context, one does not even know on which domain S and E are defined, let alone that it would be possible to grasp the significance of $\dot{S} \geqslant Q/T$.

FUTURE DEVELOPMENTS

In this section we will speculate on the influence which systems theory may have on physics and the possible research directions to which this may lead. We will first consider deterministic systems and comment afterwards on the stochastic case.

In our opinion are cannot expect systems theory to be helpful in very concrete situations. For that, systems theory is a field which is too much involved with generalities. For example, in physics an inverse square law is of great relevance but a systems theorist would undoubtedly try to imbed a dynamical equation which involves such a law in a general class of systems which exhibit some similar properties and it is therefore unlikely that one could achieve specific new conclusions by looking from this more general vantage point.

However, as we have argued, systems theory brings with it an important new axiomatic point of view from which to regard dynamical phenomena. Specifically, from the outset it considers dynamical systems with external variables, and not just autonomous systems. This starting point then leads to the state space realization problem: the representation of a given external behaviour in terms of state variables. In our opinion this axiomatic point of view and the questions which one can derive from it yield interesting research topics which may eventually influence physics, particularly on an elementary linguistic level.

A more challenging but certainly no less promising area is that of stochastic systems. An important element in such systems is the role of observations. In physics (with the exception of quantum mechanics) observations play a secondary role even when one is considering stochastic models, whereas in systems theory, especially in stochastic control, observations play a central role. The difficulty, however, is that there does not exist as yet a suitably general axiomatic framework for stochastic systems with external variables. The role of observations in such a general setting remains a particularly challenging one to clarify and to axiomatize. From such a framework, it would undoubtedly be possible to pass to questions of realization theory and make contact with statistical and quantum mechanics. It is in this context that we expect a great deal of contact between both fields (as, for example, in [7]). It is interesting to observe that the programme of 'stochastic mechanics' [8] which has as its goal the stochastic interpretation of quantum mechanics, may, in fact, be viewed as an effort in stochastic realization theory.

REFERENCES

[1] J. C. Willems, 'System Theoretic Models for the Analysis of Physical System', *Ricerche di Automatica*, Special issue on Systems Theory and Physics **X** (1979), 71–106.
[2] A. J. van der Schaft, 'Hamiltonian Dynamics with External Forces and Observations', *Mathematical Systems Theory* **15** (1982), 145–168.
[3] A. J. van der Schaft, *System Theoretic Descriptions of Physical Systems* (Mathematical Centre Tracts, Amsterdam, 1983) (in press).
[4] A. J. van der Schaft, 'Symmetries and Conservation Laws for Hamiltonian systems with Inputs and Outputs: A Generalization of Noether's Theorem', *Systems and Control Letters* **1** (1981), 108–15.
[5] J. C. Willems, 'System Theoretic Foundations for Modelling Physical Systems', in *Dynamical Systems and Microphysics* (eds. A. Blaquière, F. Fer, and A. Marzollo), CISM Lecture Notes No. 261 (Springer-Verlag, Heidelberg and New York, 1980), pp. 279–90.
[6] J. C. Willems and A. J. van der Schaft, 'Modelling of Dynamical Systems Using External and Internal Variables with Applications to Hamilton Systems', in *Dynamical Systems and Microphysics* (eds. A. Arez, A. Blaquière, and A. Marzollo) (Academic Press, New York, 1982), pp. 233–64.
[7] S. K. Mitter, 'Nonlinear Filtering and Stochastic Mechanics', in *Stochastic Systems: The Mathematics of Filtering and Identification and Applications* (eds. M. Hazewinkel and J. C. Willems) (Reidel Publishing Co., Dordrecht, (1981), pp. 479–504.
[8] F. Guerra, 'Structural Aspects of Stochastic Mechanics and Stochastic Field Theory', *Physics Reports* **77** (1981), 263–312.

Mathematics Institute, Groningen, The Netherlands

Index

GPSR Compliance

The European Union's (EU) General Product Safety Regulation (GPSR) is a set of rules that requires consumer products to be safe and our obligations to ensure this.

If you have any concerns about our products, you can contact us on ProductSafety@springernature.com

In case Publisher is established outside the EU, the EU authorized representative is:

Springer Nature Customer Service Center GmbH
Europaplatz 3
69115 Heidelberg, Germany

Batch number: 10614746

Printed by Printforce, the Netherlands